Reliability Growth

ENHANCING DEFENSE SYSTEM RELIABILITY

Panel on Reliability Growth Methods for Defense Systems

Committee on National Statistics

Division of Behavioral and Social Sciences and Education

NATIONAL RESEARCH COUNCIL
OF THE NATIONAL ACADEMIES

THE NATIONAL ACADEMIES PRESS
Washington, D.C.
www.nap.edu

THE NATIONAL ACADEMIES PRESS 500 Fifth Street, NW Washington, DC 20001

This study was supported by an award between the National Academy of Sciences and the U.S. Department of Defense through the National Science Foundation. Support for the work of the Committee on National Statistics is provided by a consortium of federal agencies through a grant from the National Science Foundation (award number SES-0453930). Any opinions, findings, conclusions, or recommendations expressed in this publication are those of the author(s) and do not necessarily reflect the view of the organizations or agencies that provided support for this project.

International Standard Book Number-13: 978-0-309-31474-9
International Standard Book Number-10: 0-309-31474-7

Additional copies of this workshop summary are available for sale from the National Academies Press, 500 Fifth Street, NW, Keck 360, Washington, DC 20001; (800) 624-6242 or (202) 334-3313; http://www.nap.edu.

Printed in the United States of America

Suggested citation: National Research Council. (2015). *Reliability Growth: Enhancing Defense System Reliability.* Panel on Reliability Growth Methods for Defense Systems, Committee on National Statistics, Division of Behavioral and Social Sciences and Education. Washington, DC: The National Academies Press.

THE NATIONAL ACADEMIES

Advisers to the Nation on Science, Engineering, and Medicine

The **National Academy of Sciences** is a private, nonprofit, self-perpetuating society of distinguished scholars engaged in scientific and engineering research, dedicated to the furtherance of science and technology and to their use for the general welfare. Upon the authority of the charter granted to it by the Congress in 1863, the Academy has a mandate that requires it to advise the federal government on scientific and technical matters. Dr. Ralph J. Cicerone is president of the National Academy of Sciences.

The **National Academy of Engineering** was established in 1964, under the charter of the National Academy of Sciences, as a parallel organization of outstanding engineers. It is autonomous in its administration and in the selection of its members, sharing with the National Academy of Sciences the responsibility for advising the federal government. The National Academy of Engineering also sponsors engineering programs aimed at meeting national needs, encourages education and research, and recognizes the superior achievements of engineers. Dr. C. D. Mote, Jr., is president of the National Academy of Engineering.

The **Institute of Medicine** was established in 1970 by the National Academy of Sciences to secure the services of eminent members of appropriate professions in the examination of policy matters pertaining to the health of the public. The Institute acts under the responsibility given to the National Academy of Sciences by its congressional charter to be an adviser to the federal government and, upon its own initiative, to identify issues of medical care, research, and education. Dr. Victor J. Dzau is president of the Institute of Medicine.

The **National Research Council** was organized by the National Academy of Sciences in 1916 to associate the broad community of science and technology with the Academy's purposes of furthering knowledge and advising the federal government. Functioning in accordance with general policies determined by the Academy, the Council has become the principal operating agency of both the National Academy of Sciences and the National Academy of Engineering in providing services to the government, the public, and the scientific and engineering communities. The Council is administered jointly by both Academies and the Institute of Medicine. Dr. Ralph J. Cicerone and Dr. C. D. Mote, Jr., are chair and vice chair, respectively, of the National Research Council.

www.national-academies.org

PANEL ON RELIABILITY GROWTH METHODS FOR DEFENSE SYSTEMS

ARTHUR FRIES (*Chair*), Institute for Defense Analyses, Alexandria, VA
W. PETER CHERRY, Science Applications International Corporation (retired), Ann Arbor, MI
ROBERT G. EASTERLING, Statistical Consultant, Cedar Crest, NM
ELSAYED A. ELSAYED, Department of Industrial and Systems Engineering, Rutgers University
APARNA V. HUZURBAZAR, Statistical Sciences Group, Los Alamos National Laboratory, Los Alamos, NM
PATRICIA A. JACOBS, Operations Research Department, Naval Postgraduate School, Monterey, CA
WILLIAM Q. MEEKER, JR., Department of Statistics, Iowa State University
NACHI NAGAPPAN, Empirical Software Engineering Group, Microsoft Research, Redmond, WA
MICHAEL PECHT, Center for Advanced Life Cycle Engineering, University of Maryland
ANANDA SEN, Department of Family Medicine, University of Michigan Health System
SCOTT VANDER WIEL, Statistical Sciences Group, Los Alamos National Laboratory, Los Alamos, NM

MICHAEL L. COHEN, *Study Director*
ERNEST SEGLIE, *Consultant*
MICHAEL J. SIRI, *Program Associate*

COMMITTEE ON NATIONAL STATISTICS
2013-2014

Acknowledgments

We first thank Frank Kendall, the Under Secretary of Defense for Acquisition, Technology, and Logistics (AT&L) and Michael Gilmore, the Director of Operational Test and Evaluation (DOT&E) for their interest in and support for this study. Over the past 20 years, these two individuals and their predecessors have provided support for a series of related projects that have produced useful studies and, equally important, helped to establish a greater degree of collaboration between the defense testing community and leading members of the statistical, system engineering, and software engineering disciplines.

We also thank our primary contacts at the U.S. Department of Defense (DoD), Nancy Spruill, director of AT&L acquisition resources and analysis, and Catherine Warner, DOT&E science advisor, who were always ready to assist the panel's work. They helped us clarify the issues for which we could have the greatest impact, they identified appropriate DoD staff to provide presentations at panel meetings to help us better understand the department's current environment and operations, and they provided us with DoD documents (e.g., handbooks, guidances, memos, etc.) relevant to our study. We are greatly indebted to these four people for their help to the panel.

The panel members are also very indebted to the many experts who provided presentations at our first three panel meetings: Darryl Ahner (AFIT/ENS), Karen T. Bain (NAVAIR), Gary Bliss (USD AT&L), Albert (Bud) Boulter (USAF SAF/AQRE), Steve Brown (Lennox), David Burdick (Boeing), Michael J. Cushing (AEC, retired), Paul Ellner (AMSAA), Michael Gilmore (DOT&E), Martha Gardner (General Electric), Don Gaver (NPGS), Jerry Gibson (ASC/ENDR), Lou Gullo (Raytheon), Brian Hall (ATEC), Frank

Kendall (USD AT&L), Shirish Kher (Alcatel-Lucent), Eric Loeb (DOT&E), Andy Long (LMI), William McCarthy (OPTEVFOR), Stephan Meschter (BAE Systems), Andy Monje (DASD SE), Ken Neubeck (Exelisinc), David Nicholls (RIAC), Paul Shedlock (Raytheon), Tom Simms (USD AT&L), Nozer Singpurwalla (GWU), Jim Streilein (DOT&E), Patrick Sul (DOT&E), Daniel Telford (AFOTEC), Nicholas Torellis (OSD), Tom Wissink (Lockheed Martin), James Woodford (ASN(RD&A)), and Guangbin Yang (Ford).

We are grateful for the help from Michael Siri on administrative arrangements, and we thank Eugenia Grohman for extremely comprehensive technical editing.

The report has been reviewed in draft form by individuals chosen for their diverse perspectives and technical expertise, in accordance with procedures approved by the Report Review Committee of the National Research Council. The purpose of this independent review is to provide candid and critical comments that assist the institution in making its reports as sound as possible and to ensure that the reports meet institutional standards for objectivity, evidence, and responsiveness to the study charge. The review comments and draft manuscript remain confidential to protect the integrity of the deliberative process.

The panel thanks the following individuals for their review of the interim report: Karen T. Bain, Reliability and Maintainability, NAVAIR; Michael J. Cushing, U.S. Army Evaluation Center (Retired); Kathleen V. Diegert, Statistics and Human Factors, Sandia; Richard T. Durrett, Math Department, Duke University; Millard S. Firebaugh, Department of Mechanical Engineering, University of Maryland, College Park; Donald P. Gaver, Jr., Operations Research, Emeritus, U.S. Naval Postgraduate School; Pradeep Lall, Department of Mechanical Engineering, University of Auburn; Paul E. Shedlock, Reliability and System Safety Department, Engineering Product Support Directorate, Raytheon Company; Neil G. Siegel, Office of the Chief Technology Officer, Northrop Grumman Information Systems; and Marlin U. Thomas, Department of Industrial and Operations Engineering, University of Michigan.

Although the reviewers listed above have provided many constructive comments and suggestions, they were not asked to endorse the conclusions or recommendations, nor did they see the final draft of the report before its release. The review of the report was overseen by Thom J. Hodgson, Fitts Industrial and Systems Engineering Department, North Carolina State University and Roderick J. Little, Department of Biostatistics, School of Public Health, University of Michigan. Appointed by the National Research Council, they were responsible for making certain that the independent examination of this report was carried out in accordance with institutional procedures and that all review comments were carefully considered. Responsibility for the final content of the report rests entirely with the authoring committee and the National Research Council. We also thank Ali Mosleh, Henry Samueli School of Engineering and Applied Science, University of California, Los Angeles, for his review of Appendix D. Finally, we thank the panel members who drafted large sections of this report and who devoted a great deal of time, energy, and expertise to this effort and always found time to contribute to the work.

Arthur Fries, *Chair*
Michael L. Cohen, *Study Director*
Panel on Reliability Growth Methods
for Defense Systems

Contents

Summary

Reliability—the innate capability of a system to perform its intended functions—is one of the key performance attributes that is tracked during U.S. Department of Defense (DoD) acquisition processes. Although every system is supposed to achieve a specified reliability requirement before being approved for acquisition, the perceived urgency to operationally deploy new technologies and military capabilities often leads to defense systems being fielded without having demonstrated adequate reliability. Between 2006 and 2011, one-half of the 52 major defense systems reported on by the DoD Office of the Director, Operational Test and Evaluation (DOT&E) to Congress failed to meet their prescribed reliability thresholds, yet all of the systems proceeded to full-rate production status.

Defense systems that fail to meet their reliability requirements are not only less likely to successfully carry out their intended missions, but also may endanger the lives of the Armed Service personnel who are depending on them. Such deficient systems are also much more likely than reliable systems to require extra scheduled and unscheduled maintenance and to demand more spare and replacement parts over their life cycles. In addition, the consequences of not finding fundamental flaws in a system's design until after it is deployed can include costly and strategic delays until expensive redesigns are formulated and implemented and imposition of operational limits that constrain tactical employment profiles.

Recognizing these costs, the Office of the Secretary of Defense (OSD)—through DOT&E and the Office of the Under Secretary of Defense for Acquisition, Technology, and Logistics (USD AT&L)—in 2008 initiated a concerted effort to elevate the importance of reliability through greater use of design-for-reliability techniques, reliability growth testing, and formal reliability growth modeling. To this end, handbooks, guidance, and formal memoranda were revised or newly issued to provide policy to lead to the reduction of the frequency of reliability deficiencies. To evaluate the efficacy of that effort and, more generally, to assess how current DoD principles and practices could be strengthened to increase the likelihood of defense systems satisfying their reliability requirements, DOT&E and USD AT&L requested that the National Research Council conduct a study through its Committee on National Statistics (CNSTAT). The Panel on Reliability Growth Methods for Defense Systems was created to carry out that study.

SCOPE AND CONTEXT

The panel examined four broad topics: (1) the processes governing the generation of reliability requirements for envisioned systems, the issuance of requests for proposals (RFPs) for new defense acquisitions, and the contents of and evaluation of proposals in response; (2) modern design for reliability and how it should be utilized by contractors; (3) contemporary reliability test and evaluation practices and how they should be incorporated into contractor and government planning and testing; and (4) the current state of formal reliability growth modeling, what functions is it useful for, and what constitutes suitable use.

The current environment for defense system acquisition differs from the conditions that prevailed in DoD in the 1990s and also differs from the circumstances faced by commercial companies. Compared to the past, today's DoD systems typically entail: greater design complexities (e.g., comprising dozens of subsystems with associated integration and interoperability issues); more dependence on software components; increased reliance on integrated circuit technologies; and more intricate dependencies on convoluted nonmilitary supply chains.

In commercial system development, all elements of program control are generally concentrated in a single project manager driven by a clear profit motive. In contrast, DoD acquisition processes are spearheaded by numerous independent "agents"—a system developer, one or more contractors and subcontractors, a DoD program manager, DoD testers, OSD oversight offices, and the military users—all of whom view acquisition from different perspectives and incentive structures. In addition, in the commercial sector the risk of delivering a poor reliability system is borne primarily by the manufacturer (in terms of reduced current and future sales, warranty costs, etc.), but for defense systems, the government and the military users generally assume most of the risk because the government is committed to extensive purchase quantities prior to the point where reliability deficiencies are evident.

Over the past few decades, commercial industries have developed two basic approaches to producing highly reliable system designs: techniques germane to the initial design, referred to as design-for-reliability methods; and testing in development phases aimed at finding failure modes and implementing appropriate design improvements to increase system reliability. In contrast, DoD has generally relied on extensive system-level testing, which is both time and cost intensive, to raise initial reliabilities ultimately to the vicinity of prescribed final reliability requirements. To monitor this growth in reliability, reliability targets are established at various intermediate stages of system developmental testing (DT). Upon the completion of DT, operational testing (OT) is conducted to examine reliability performance under realistic conditions with typical military users and maintainers. The recent experience with this DoD system development strategy is that operational reliability has frequently been deficient, and that deficiency can generally be traced back to reliability shortfalls in the earliest stages of DT.

Central to current DoD approaches to reliability are reliability growth models, which are mathematical abstractions that explicitly link expected gains in system reliability to total accrued testing time. They facilitate the design of defensible reliability growth testing programs and they support the tracking of the current system reliability. As is true for modeling in general, applications of reliability growth models entail implicit conceptual assumptions whose validity needs to be independently corroborated.

DoD reliability testing, unless appropriately modulated, does not always align with the theoretical underpinnings of reliability growth formulations, such as that system operating circumstances (i.e., physical environments, stresses that test articles are subjected to, and potential failure modes) do not vary during reliability growth periods.

The common interpretation of the term "reliability" has broad ramifications throughout DoD acquisition, from the statement of performance requirements to the demonstration of reliability in operational testing and evaluation. Because requirements are prescribed well in advance of testing, straightforward articulations, such as mean-time-between failures (MTBF) and probability of success, are reasonable. Very often, the same standard MTBF and success probability metrics will be appropriate for describing established levels of system reliability for the data from limited duration testing. But there may be instances—depending on sample sizes, testing conditions, and test prototypes—for which more elaborate analysis and reporting methods would be appropriate. More broadly, system reliabilities, both actual and estimated, reflect the particulars of testing circumstances, and these circumstances may not match intended operational usage profiles.

PANEL OBSERVATIONS AND RECOMMENDATIONS

The Panel on Reliability Growth Methods for Defense Systems offers 25 recommendations for improving the reliability of U.S. defense systems. These are listed in entirety at the end of this Executive Summary and are discussed in detail within the body of this report. Here we first summarize the panel's primary observations that underlie the resultant recommendations. Then we highlight the content and substance of the individual recommendations. The panel's conclusions cover the entire spectrum of DoD acquisition activities:

- DoD has taken a number of essential steps toward developing systems that satisfy prescribed operational reliability requirements and perform dependably once deployed.
- Fundamental elements of reliability improvement should continue to be emphasized, covering:
  —operationally meaningful and attainable requirements;
  —requests for proposal and contracting procedures that give prominence to reliability concerns;
  —design-for-reliability activities that elevate the level of initial system reliability;
  —focused test and evaluation events that grow system reliability and provide comprehensive examinations of operational reliability;
  —appropriate applications of reliability growth methodologies (i.e., compatible with underlying assumptions) for determining the extent of system-level reliability testing and the validity of assessment results;
  —empowered hardware and software reliability management teams that direct contractor design and test activities;
  —feedback mechanisms, spanning reliability design, testing, enhancement initiatives, and postdeployment performance, that inform current and future developmental programs; and
  —DoD review and oversight processes.
- Sustained funding is needed throughout system definition, design, and development, to:
  —incentivize contractor reliability initiatives;
  —accommodate planned reliability design and testing activities, including any revisions that may arise; and
  —provide sufficient state-of-the-art expertise to support DoD review and oversight.

Support for the recommendations that are put forward is provided throughout the report, and they are further discussed and presented in the final chapter. Here we present the content of the recommendations in terms of four aspects of the acquisition process: (1) system requirements, RFPs, and proposals; (2) design for reliability; (3) reliability testing and evaluation; and (4) reliability growth models.

The recommendations include a few "repeats"—endorsements of earlier CNSTAT and DoD studies, as well as reformulations of existing DoD acquisition procedures and regulations. These are presented to provide a complete self-contained rendition of reliability enhancement proposals, and because current DoD guidance and governance have not been fully absorbed, are inconsistently applied, and are subject to change.

System Requirements, RFPs, and Proposals

Prior to the initiation of a defense acquisition program, the performance requirements of the planned system, including reliability, have to be formally established. The reliability requirement should be grounded in terms of operational relevance (e.g., mission success) and be linked explicitly (within the fidelity available at this early stage) to the costs of acquisition and sustainment over the lifetime of the system. This operational reliability requirement also has to be technically feasible (i.e., verified to be within the state-of-the-art of current or anticipated near-term scientific, engineering, and manufacturing capabilities). Finally, the operational reliability requirement needs to be measureable and testable. The process for developing the system reliability requirement should draw on pertinent previous program histories and use the resources in OSD and the services (including user and testing communities). Steps should be reviewed and supplemented, as needed, by external subject-matter experts with reliability engineering and other technical proficiencies relevant to the subject system. **[Recommendations 1, 2, 24, and 25]**

The reliability requirement should be designated as a key performance parameter, making compliance contractually mandatory. This designation would emphasize the importance of reliability in the acquisition process and enhance the prospects of achieving suitable system reliability. During developmental testing, opportunities to relax the reliability requirement should be limited: it should be permitted only after high-level review and approval (at the level of a component acquisition authority or higher), and only after studying the potential effects on mission accomplishment and life-cycle costs. **[Recommendations 3 and 5]**

The government's RFP should contain sufficient detail for contractors to specify how they would design, test, develop, and qualify the envisioned system and at what cost levels. The RFP needs to elaborate on reliability requirements and justifications, hardware and software considerations, operational performance profiles and circumstances, anticipated environmental load conditions, and definitions of "system failure." The preliminary versions of the government's concept for a phased developmental testing program (i.e., timing, size, and characteristics of individual testing events) should also be provided. The government's evaluations of contractor proposals should consider the totality of the proffered reliability design, testing, and management processes, including specific failure definitions and scoring criteria to be used for contractual verification at various intermediate system development points. **[Recommendations 1, 2, 4, 7, and 16]**

Design for Reliability

High reliability early in system design is better than extensive and expensive system-level developmental testing to correct low initial reliability levels. The former has been the common *successful* strategy in non-DoD commercial acquisition; the latter has been the predominantly *unsuccessful* strategy in DoD acquisition.

Modern design-for-reliability techniques include but are not limited to: (1) failure modes and effects analysis, (2) robust parameter design, (3) block diagrams and fault tree analyses, (4) physics-of-failure methods, (5) simulation methods, and (6) root-cause analysis. The appropriate mix of methods will vary across systems. At the preliminary stages of design, contractors should be able to build on the details offered in RFPs, subsequent government interactions, and past experience with similar types of systems. **[Recommendation 6]**

The design process itself should rest on appropriately tailored applications of sound reliability engineering practices. It needs not only to encompass the intrinsic hardware and software characteristics of system performance, but also to address broader reliability aspects anticipated for manufacturing, assembly, shipping and handling, life-cycle profiles, operation, wear-out and aging, and maintenance and repair. Most importantly, it has to be supported by a formal reliability management structure and adequate funding (possibly including incentives) that provides for the attainment and demonstration of high reliability levels early in a system's design and development phases. If a system (or one or more of its subsystems) is software intensive, then the contractor should be required to provide a rationale for its selection of a software architecture and management plan, and that plan should be reviewed by independent subject-matter experts appointed by DoD. Any major changes made after the initial system design should be assessed for their potential impact on subsequent design and testing activities, and the associated funding needs should be provided to DoD. **[Recommendations 6, 7, 15, and 18]**

Three specific aspects of design for reliability warrant emphasis. First, more accurate predictions of reliabilities for electronic components are needed. The use of Military Handbook (MIL-HDBK) 217 and its progeny have been discredited as being invalid and inaccurate: they should be replaced with physics-of-failure methods and with estimates based on validated models. Second, software-intensive systems and subsystems merit special scrutiny, beginning in the early conceptual stages of system design. A contractor's development of the software architecture, specifications, and oversight management plan need to be reviewed independently by DoD and external subject-matter experts in software reliability engineering. Third, holistic design methods should be pursued to address hardware, software, and human factors elements of system reliability—not as compartmentalized concerns, but via integrated approaches that comprehensively address potential interaction failure modes. **[Recommendations 6, 8, and 9]**

Reliability Testing and Evaluation

Increasing reliability after the initial system design is finalized involves interrelated steps in planning for acquiring system performance information through testing, conducting various testing events, evaluating test results, and iteration. There are no universally applicable algorithms that precisely prescribe the composition and sequencing of individual activities for software and hardware developmental testing and evaluation at the component, subsystem, and system levels. General principles and strategies, of which we are broadly supportive, have been espoused in a number of recent documents introduced to and utilized by various segments of DoD acquisition communities. While the reliability design and testing topics addressed in these documents are extensive, the presented expositions are not in-depth and applications to specific acquisition programs have to draw upon seasoned expertise in a number of reliability domains—reliability engineering, software reliability engineering, reliability modeling, accelerated testing, and the reliability of electronic components. In each of these domains, DoD needs to add appropriate proficiencies through combinations of in-house hiring, consulting or contractual agreements, and training of current personnel.

DoD also needs to develop additional expertise in advances in the state-of-the-art of reliability practices to respond to challenges posed by technological complexities and by endemic schedule and budget constraints. Innovations should be pursued in several domains: the foundations of design for reliability; early developmental testing and evaluation (especially for new technologies and for linkages to physical failure mechanisms); planning for efficient testing and evaluation and comprehensive data assimilation (for different classes of defense systems); and techniques for assessing aspects of near- and long-term reliability that are not well-addressed in dedicated testing.

Finally, to promote learning, DoD should encourage the establishment of information-sharing repositories that document individual reliability program histories (e.g., specific design and testing and evaluation initiatives) and demonstrated reliability results from developmental and operational testing and evaluation and postdeployment operation. Also needed are descriptions of system operating conditions, as well as manufacturing methods and quality controls, component suppliers, material and design changes, and other relevant information. This database should be used to inform additional acquisitions of the same system and for planning and conducting future acquisition programs of related systems. In developing and using this database, DoD needs to ensure that the data are fully protected against the disclosure of proprietary and classified information. **[Recommendations 22, 23, 24, and 25]**

Planning for and conducting a robust testing program that increases system reliability, both hardware and software, requires that sufficient funds be allocated for testing and oversight of contractor and subcontractor activities. Such funding needs to be dedicated exclusively to testing so that it cannot be later redirected for other purposes. The amount of such funding needs to be a consideration in making decisions about proposals, in awarding contracts, and in setting incentives for contractors. The execution of a developer's reliability testing program should be overseen and governed by a formal reliability management structure that is empowered to make reliability an acquisition priority (beginning with system design options), retains flexibility to respond to emerging insights and observations, and comprehensively archives hardware and software reliability testing, data, and assessments. Complete documentation should be budgeted for and made available to all relevant program and DoD entities. **[Recommendations 6, 7, 9, 12, 15, 16, 17, and 18]**

The government and contractor should collaborate to further develop the initial developmental testing and evaluation program for reliability outlined in the RFP and described in the contractor's proposal. Reliability test plans, both hardware and software, should be regularly reviewed (by DoD and the developer) and updated as needed (e.g., at major design reviews)—considering what has been demonstrated to date about the attainment of reliability goals, contractual requirements, and intermediate thresholds and what remains uncertain about component, subsystem, and system reliability. Interpretations should be cognizant of testing conditions and how they might differ from operationally realistic circumstances. **[Recommendations 4, 7, and 11]**

The objectives for early reliability developmental testing and evaluation, focused at the component and subsystem levels, should be to surface failure mechanisms, inform design enhancement initiatives, and support reliability assessments. The scope for these activities, for both hardware and software systems, should provide timely assurance that system reliability is on track with expectations. The goal should be to identify and address substantive reliability deficiencies at this stage of development, when they are least costly, before designs are finalized and system-level production is initiated.

For hardware components and subsystems, there are numerous "accelerated" testing approaches available to identify, characterize, and assess failure mechanisms and reliability within the limited time afforded in early developmental testing and evaluation. They include exposing test articles to controlled nonstandard overstress environments and invoking physically plausible models to translate observed results to nominal use conditions. To manage software development in this early phase, contractors should be required to test the full spectrum of usage profiles, implement meaningful performance metrics to track software completeness and maturity, and chronicle results. For software-intensive systems and subsystems, contractors should be required to develop automated software testing tools and supporting documentation and to provide these for review by an outside panel of subject-matter experts appointed by DoD. **[Recommendations 7, 9, 12, and 14]**

When system prototypes (or actual systems) are produced, system-level reliability testing can begin, but that should not occur until the contractor offers a statistically supportable estimate of the current system reliability that is compatible with the starting system reliability requirement prescribed in the program's reliability demonstration plan. System-level reliability testing typically proceeds, and should proceed, in discrete phases, interspersed by corrective action periods in which observed failure modes are assessed, potential design enhancements are postulated, and specific design improvements are implemented. Individual test phases should be used to explore system performance capabilities under different combinations of environmental and operational factors and to demonstrate levels of achieved reliability specific to the conditions of that test phase (which may or may not coincide precisely with operationally realistic scenarios). Exhibited reliabilities, derived from prescribed definitions of system hardware and software failures, should be monitored and tracked against target reliabilities to gauge progress toward achieving the formal operational reliability requirement. Of critical importance is the scored reliability at the beginning of system-level developmental testing, which is a direct reflection of the quality of the system design and production processes. A common characteristic of recent reliability deficient DoD programs has been early evidence of demonstratively excessive observed failure counts, especially within the first phase of reliability testing. **[Recommendations 7 and 19]**

Inadequate system-level developmental testing and evaluation results in imprecise or misleading direct assessments of system reliability. If model-based estimates (e.g., based on accelerated testing of major subsystems) become integral to demonstrating achieved system reliability and supporting major acquisition decisions, then the modeling should be subject to review by an independent panel of appointed subject-matter experts. To enhance the prospects of growing operational reliability, developmental system-level testing should incorporate elements of operational realism to the extent feasible. At a minimum, a single full-system, operationally relevant developmental test event should be scheduled near the end of developmental testing and evaluation—with advancement to operational testing and evaluation contingent on satisfaction of the system operational reliability requirement or other justification (e.g., combination of proximate reliability estimate, well-understood failure modes, and tenable design improvements). **[Recommendations 13 and 20]**

In operational testing, each event ideally would be of a sufficiently long duration to provide a stand-alone statistically defensible assessment of the system's operational reliability for distinct operational scenarios and usage conditions. When operational testing and evaluation is constrained (e.g., test hours or sample sizes are limited) or there are questions of interpretation (e.g., performance heterogeneity across test articles or operational factors is detected), nonstandard sophisticated analyses may be required to properly characterize the system's operational reliability for a single test event or synthesizing data from multiple developmental and operational test events. Follow-on operational testing and evaluation may be required to settle unresolved issues, and DoD should ensure that it is done. If the attainment of an adequate level of system operational reliability has not been demonstrated with satisfactory confidence, then DoD should not approve the system for full-rate production and fielding without a formal review of the likely effects that the deficient reliability will have on the probability of mission success and system life-cycle costs. **[Recommendation 21]**

The glimpses of operational reliability offered by operational testing are not well suited for identifying problems that relate to longer use, such as material fatigue, environmental effects, and aging. These considerations should be addressed in the design phase and in developmental testing and evaluation (using accelerated testing), and their manifestations should be recorded in the postdeployment reliability history database established for the system. **[Recommendation 22]**

Reliability Growth Models

DoD applications of reliability growth models, focused on test program planning and reliability data assessments, generally invoke a small number of common analytically tractable constructs. The literature, however, is replete with other viable formulations—for time-to-failure data and discrete success/failure and both hardware and software systems (code). No particular reliability growth model is universally dominant for all potential applications, and some data complexities demand that common modeling approaches be modified in nonstandard and novel ways. **[Recommendations 10, 11, and 19]**

Within current formal DoD test planning documentation, each developmental system is required to establish an initial reliability growth curve (i.e., graphical depiction of how system reliability is planned to increase over the allotted developmental period) and to revise the curve as needed when program milestones are achieved or in response to unanticipated testing outcomes. The curve can be constructed from applying a reliability growth model, incorporating historical precedence from previous developmental programs, or customizing hybrid approaches. It should be fully integrated with overall system developmental test and evaluation strategies (e.g., accommodating other nonreliability performance issues) and retain adequate flexibility to respond to emerging testing results—while recognizing potential sensitivities to underlying analytical assumptions. The strategy of building the reliability growth curve to bring the system operational reliability at the end of developmental test and evaluation to a reasonable point supporting the execution of a stand-alone operational test and evaluation, with acceptable statistical performance characteristics, is eminently reasonable. Some judgment will always be needed in determining the number, size, and composition of individual developmental testing events, accounting for the commonly experienced DT/OT reliability gap, and in balancing developmental and operational testing and evaluation needs with schedule and funding constraints. **[Recommendations 10 and 11]**

Reliability growth models can be used, when supporting assumptions hold, as plausible "curve fitting" mechanisms for matching observed test results to prescribed model formulations—for tracking the development and maturity of software in early developmental testing, and for tracking the progression of system reliability during system-level testing. When overall sample sizes (i.e., numbers of recorded failures across multiple tests) are large, modeling can enhance the statistical precision associated with the last test event and support program oversight judgments. No elaborate modeling is needed, however, when the initial developmental testing experiences far more failures than anticipated by the planned reliability growth trajectory—indicative of severe reliability design deficiencies. **[Recommendations 9, 10, and 19]**

Standard applications of common reliability growth methods can yield misleading results when some test events are more stressful than others, when system operating profiles vary across individual tests, or when system functionality is added incrementally over the course of developmental testing. Under such nonhomogeneous circumstances, tenable modeling may need to require the development and validation of separate reliability growth models for distinct components of system reliability, flexible regression-based formulations, or other sophisticated analytical approaches. Without adequate data, however, more complex models can be difficult to validate: in this circumstance, too, reliability growth modeling needs to recognize the limitations of trying to apply sophisticated statistical techniques to the data. The utility and robustness of alternative specifications of reliability growth models and accompanying statistical methodologies can be explored via simulation studies. The general caution against model-based extrapolations outside of the range of the supporting test data applies to projections of observed patterns of system reliability growth to future points in time. One important exception, from a program oversight perspective, is assessing the reliability growth potential when a system clearly is experiencing reliability shortfalls during developmental testing—far below initial target values or persistently less than a series of goals. Reliability growth methods, incorporating data on specific exhibited failure modes and the particulars of testing circumstances, can demonstrate that there is little chance for the program to succeed unless major system redesigns and institutional reliability management improvements are implemented (i.e., essentially constituting a new reliability growth program). **[Recommendations 10 and 19]**

LIST OF RECOMMENDATIONS

RECOMMENDATION 1 **The Under Secretary of Defense for Acquisition, Technology, and Logistics should ensure that all analyses of alternatives include an assessment of the relationships between system reliability and mission success and between system reliability and life-cycle costs.**

RECOMMENDATION 2 **Prior to issuing a request for proposal (RFP), the Office of the Under Secretary of Defense for Acquisition, Technology, and Logistics should issue a technical report on the reliability requirements and their associated justification. This report should include the estimated relationship between system reliability and total acquisition and life-cycle costs and the technical justification that the reliability requirements for the proposed new system are feasible, measurable, and testable. Prior to being issued, this document should be reviewed by a panel with expertise in reliability engineering, with members from the user community, from the testing community, and from outside of the service assigned to the acquisition. We recognize that before any development has taken place these assessments are somewhat guesswork and it is the expectation that as more about the system is determined, the assessments can be improved. Reliability** engineers of the services involved in each particular acquisition should have full access to the technical report and should be consulted prior to the finalization of the RFP.

RECOMMENDATION 3 Any proposed changes to reliability requirements by a program should be approved at levels no lower than that of the service component acquisition authority. Such approval should consider the impact of any reliability changes on the probability of successful mission completion as well as on life-cycle costs.

RECOMMENDATION 4 Prior to issuing a request for proposal (RFP), the Under Secretary of Defense for Acquisition, Technology, and Logistics should mandate the preparation of an outline reliability demonstration plan that covers how the department will test a system to support and evaluate system reliability growth. The description of these tests should include the technical basis that will be used to determine the number of replications and associated test conditions and how failures are defined. The outline reliability demonstration plan should also provide the technical basis for how test and evaluation will track in a statistically defendable way the current reliability of a system in development given the likely number of government test events as part of developmental and operational testing. Prior to being included in the request for proposal for an acquisition program, the outline reliability demonstration plan should be reviewed by an expert external panel. Reliability engineers of the services involved in the acquisition in question should also have full access to the reliability demonstration plan and should be consulted prior to its finalization.

RECOMMENDATION 5 The Under Secretary of Defense for Acquisition, Technology, and Logistics should ensure that reliability is a key performance parameter: that is, it should be a mandatory contractual requirement in defense acquisition programs.

RECOMMENDATION 6 The Under Secretary of Defense for Acquisition, Technology, and Logistics should mandate that all proposals specify the design-for-reliability techniques that the contractor will use during the design of the system for both hardware and software. The proposal budget should have a line item for the cost of design-for-reliability techniques, the associated application of reliability engineering methods, and schedule adherence.

RECOMMENDATION 7 The Under Secretary of Defense for Acquisition, Technology, and Logistics should mandate that all proposals include an initial plan for system reliability and qualification (including failure definitions and scoring criteria that will be used for contractual verification), as well as a description of their reliability organization and reporting structure. Once a contract is awarded, the plan should be regularly updated, presumably at major design reviews, establishing a living document that contains an up-to-date assessment of what is known by the contractor about hardware and software reliability at the component, subsystem, and system levels. The U.S. Department of Defense should have access to this plan, its updates, and all the associated data and analyses integral to their development.

RECOMMENDATION 8 Military system developers should use modern design-for-reliability (DFR) techniques, particularly physics-of-failure (PoF)-based methods, to support system design and reliability estimation. MIL-HDBK-217 and its progeny have grave deficiencies; rather, the U.S. Department of Defense should emphasize DFR and PoF implementations when reviewing proposals and reliability program documentation.

RECOMMENDATION 9 For the acquisition of systems and subsystems that are software intensive, the Under Secretary of Defense for Acquisition, Technology, and Logistics should ensure that all proposals specify a management plan for software development and also mandate that, starting early in development and continuing throughout development, the contractor provide the U.S. Department of Defense with full access to the software architecture, the software metrics being tracked, and an archived log of the management of system development, including all failure reports, time of their incidence, and time of their resolution.

RECOMMENDATION 10 The validity of the assumptions underlying the application of reliability growth models should be carefully assessed. In cases where such validity remains in question: (1) important decisions should consider the sensitivity of results to alternative model formulations and (2) reliability growth models should not be used to forecast substantially into the future. An exception to this is early in system development, when reliability growth models, incorporating relevant historical data, can be invoked to help scope the size and design of the developmental testing programs.

RECOMMENDATION 11 The Under Secretary of Defense for Acquisition, Technology, and Logistics should mandate that all proposals obligate the contractor to specify an initial reliability growth plan and the outline of a testing program to support it, while recognizing that both of these constructs are preliminary and will be modified through development. The required plan will include, at a minimum, information on whether each test is a test of components, of subsystems, or of the full system; the scheduled dates; the test design; the test scenario conditions; and the number of replications in each scenario. If a test is an accelerated test, then the acceleration factors need to be described. The contractor's budget and master schedules should be required to contain line items for the cost and time of the specified testing program.

RECOMMENDATION 12 The Under Secretary of Defense for Acquisition, Technology, and Logistics should mandate that contractors archive and deliver to the U.S. Department of Defense (DoD), including to the relevant operational test agencies, all data from reliability testing and other analyses relevant to reliability (e.g., modeling and simulation) that are conducted. This should be comprehensive and include data from all relevant assessments, including the frequency under which components fail quality tests at any point in the production process, the frequency of defects from screenings, the frequency of defects from functional testing, and failures in which a root-cause analysis was unsuccessful (e.g., the frequency of instances of failure to duplicate, no fault found, retest OK). It should also include all failure reports, times of failure occurrence, and times of failure resolution. The budget for acquisition contracts should include a line item to provide DoD with full access to such data and other analyses.

RECOMMENDATION 13 The Office of the Secretary of Defense for Acquisition, Technology, and Logistics, or, when appropriate, the relevant service program executive office, should enlist independent external, expert panels to review (1) proposed designs of developmental test plans critically reliant on accelerated life testing or accelerated degradation testing and (2) the results and interpretations of such testing. Such reviews should be undertaken when accelerated testing inference is of more than peripheral importance—for example, if applied at the major subsystem or system level, there is inadequate corroboration provided by limited system testing, and the results are central to decision making on system promotion.

RECOMMENDATION 14 For all software systems and subsystems, the Under Secretary of Defense for Acquisition, Technology, and Logistics should mandate that the contractor provide the U.S. Department of Defense (DoD) with access to automated software testing capabilities to enable DoD to conduct its own automated testing of software systems and subsystems.

RECOMMENDATION 15 The Under Secretary of Defense for Acquisition, Technology, and Logistics should mandate the assessment of the impact of any major changes to system design on the existing plans for design-for-reliability activities and plans for reliability testing. Any related proposed changes in fund allocation for such activities should also be provided to the U.S. Department of Defense.

RECOMMENDATION 16 The Under Secretary of Defense for Acquisition, Technology, and Logistics should mandate that contractors specify to their subcontractors the range of anticipated environmental load conditions that components need to withstand.

RECOMMENDATION 17 The Under Secretary of Defense for Acquisition, Technology, and Logistics should ensure that there is a line item in all acquisition budgets for oversight of subcontractors' compliance with reliability requirements and that such oversight plans are included in all proposals.

RECOMMENDATION 18 The Under Secretary of Defense for Acquisition, Technology, and Logistics should mandate that proposals for acquisition contracts include appropriate funding for design-for-reliability activities and for contractor testing in support of reliability growth. It should be made clear that the awarding of contracts will include consideration of such fund allocations. Any changes to such allocations after a contract award should consider the impact on probability of mission success and on life-cycle costs, and at the minimum, require approval at the level of the service component acquisition authority.

RECOMMENDATION 19 The Under Secretary of Defense for Acquisition, Technology, and Logistics should mandate that prior to delivery of prototypes to the U.S. Department of Defense for developmental testing, the contractor must provide test data supporting a statistically valid estimate of system reliability that is consistent with the operational reliability requirement. The necessity for this should be included in all requests for proposals.

RECOMMENDATION 20 Near the end of developmental testing, the Under Secretary of Defense for Acquisition, Technology, and Logistics should mandate the use of a full-system, operationally relevant developmental test during which the reliability performance of the system will equal or exceed the required levels. If such performance is not achieved, then justification should be required to support promotion of the system to operational testing.

RECOMMENDATION 21 The U.S. Department of Defense should not pass a system that has deficient reliability to the field without a formal review of the resulting impacts the deficient reliability will have on the probability of mission success and system life-cycle costs.

RECOMMENDATION 22 The Under Secretary of Defense for Acquisition, Technology, and Logistics should emplace acquisition policies and programs that direct the services to provide for the collection and analysis of postdeployment reliability data for all fielded systems, and to make that data available to support contractor closed-loop failure mitigation processes. The collection and analysis of such data should be required to include defined, specific feedback about reliability problems surfaced in the field in relation to manufacturing quality controls and indicate measures taken to respond to such reliability problems. In addition, the contractor should be required to implement a comprehensive failure reporting, analysis and corrective action system that encompasses all failures (regardless whether failed items are restored/repaired/replaced by a different party, e.g., subcontractor or original equipment manufacturer).

RECOMMENDATION 23 After a system is in production, changes in component suppliers or any substantial changes in manufacturing and assembly, storage, shipping and handling, operation, maintenance, and repair should not be undertaken without appropriate review and approval. Reviews should be conducted by external expert panels and should focus on impact on system reliability. Approval authority should reside with the program executive office or the program manager, as determined by the U.S. Department of Defense. Approval for any proposed change should be contingent upon certification that the change will not have a substantial negative impact on system reliability or a formal waiver explicitly documenting justification for such a change.

RECOMMENDATION 24 The Under Secretary of Defense for Acquisition, Technology, and Logistics should create a database that includes three elements obtained from the program manager prior to government testing and from the operational test agencies when formal developmental and operational tests are conducted: (1) outputs, defined as the reliability levels attained at various stages of development; (2) inputs, defined as the variables that describe the system and the testing conditions; and (3) the system development processes used, that is, the reliability design and reliability testing specifics. The collection of these data should be carried out separately for major subsystems, especially software subsystems.

RECOMMENDATION 25 To help provide technical oversight regarding the reliability of defense systems in development, specifically, to help develop reliability requirements, to review acquisition proposals and contracts regarding system reliability, and to monitor acquisition programs through development, involving the use of design-for-reliability methods and reliability testing, the U.S. Department of Defense should acquire, through in-house hiring, through consulting or contractual agreements, or by providing additional training to existing personnel, greater access to expertise in these five areas: (1) reliability engineering, (2) software reliability engineering, (3) reliability modeling, (4) accelerated testing, and (5) the reliability of electronic components.

1

Introduction

Fielded defense systems that fail to meet their reliability goals or requirements reduce the effectiveness and safety of the system and incur costs that generally require funds to be diverted from other defense needs. This is not a new problem, as readers of this report likely know. A synopsis of the relevant history is presented in the second section of this chapter. In recognition of this continuing problem, the U.S. Department of Defense (DoD) asked the National Research Council, through its Committee on National Statistics, to undertake a study on reliability.

PANEL CHARGE AND SCOPE OF STUDY

DoD originally asked the Panel on Reliability Growth Methods for Defense Systems to provide an assessment only of the use of reliability growth models to address a portion of the problem. Reliability growth models are used to track the extent to which the reliability of a system in development is on a trajectory that is consistent with achieving the system requirement by the time of its anticipated promotion to full-rate production. However, the importance of the larger problem of the failure of defense systems to achieve required reliability levels resulted in the broadening of the panel's charge. The sponsor and the panel recognized that reliability growth is more than a set of statistical models applied to testing histories. Reliability is grown through development of reasonable requirements, through design, through engineering, and through testing. Thus, DoD broadened its charge to the panel to include recommendations for procedures and techniques to improve system reliability during the acquisition process:[1]

> The present project on reliability growth methods is the latest in a series of studies at the National Research Council (NRC) on improving the statistical methods used in defense systems development and acquisition. It features a public workshop with an agenda that will explore ways in which reliability growth processes (including design, testing, and management activities) and dedicated analysis models (including estimation, tracking, and prediction methodologies) can be used to improve the development and operational performance of defense systems. Through invited presentations and discussion, the workshop will characterize commonly used and potentially applicable reliability growth methods for their suitability to defense acquisition. The scope of the workshop and list of program participants will be developed by an expert ad hoc panel that will also write the final report that summarizes the findings of the workshop and provides recommendations to the U.S. Department of Defense.

In response, the panel examined the full process of design, testing, and analysis. We began with the requested workshop that featured DoD, contractor, academic, and commercial perspectives on the issue of reliability growth methods; see Appendix B for the workshop agenda and list of participants. And, as noted in the charge, this report builds on previous work by the NRC's Committee on National Statistics.

The procedures and techniques that can be applied during system design and development include system design techniques that explicitly address reliability and testing focused on reliability improvement. We also consider when and how reliability growth models can be used to assess and track system reliability during development and in the field. In addition, given the broad mandate from the sponsor, we examined four other topics:

1. the process by which reliability requirements are developed
2. the contents of acquisition requests for proposals that are relevant to reliability
3. the contents of the resulting proposals that are relevant to reliability, and
4. the contents of acquisition contracts that are relevant to reliability.

Broadly stated, we argue throughout this report that DoD has to give higher priority to the development of reliable systems throughout all phases of the system acquisition process. This does not necessarily mean additional

---

[1]For a list of the findings and recommendations from the previous reports noted in the first sentence of the charge, see Appendix A.

funds, because in many cases what is paid for up front to improve reliability can be recovered multiple times by reducing system life-cycle costs. This latter point is supported by a U.S. Government Accountability Office report (2008, p. 7), which found that many programs had encountered substantial reliability problems in development or after initial fielding:

> Although major defense contractors have adopted commercial quality standards in recent years, quality and reliability problems persist in DOD weapon systems. Of the 11 weapon systems GAO reviewed, these problems have resulted in billions of dollars in cost overruns, years of schedule delays, and reduced weapon system availability. Prime contractors' poor systems engineering practices related to requirements analysis, design, and testing were key contributors to these quality problems.

The report further noted (U.S. Government Accountability Office, 2008, p. 19):

> [I]n DOD's environment, reliability is not usually emphasized when a program begins, which forces the department to fund more costly redesign or retrofit activities when reliability problems surface later in development or after a system is fielded. The F-22A program illustrates this point. Because DOD as the customer assumed most of the financial risk on the program, it made the decision that system development resources primarily should be focused on requirements other than reliability, leading to costly quality problems. After seven years in production, the Air Force had to budget an additional unplanned $400 million for the F-22A to address numerous quality problems and help the system achieve its baseline reliability requirements.

ACHIEVING RELIABILITY REQUIREMENTS: KEY HISTORY AND ISSUES

The magnitude of the problem in achieving reliability requirements was described at the panel's workshop by Michael Gilmore, the Director of Operational Test and Evaluation (DOT&E), and by Frank Kendall, Acting Under Secretary of Defense for Acquisition, Technology, and Logistics (USD AT&L). Since 1985, 30 percent of 170 systems under the purview of DOT&E had been reported as not having demonstrated their reliability requirements. A separate review by DOT&E in fiscal 2011 found that 9 of 15 systems, 60 percent, failed to meet their reliability thresholds. Figures for the preceding 3 years, 2008-2011, as documented in DOT&E Annual Reports, were 46 percent, 34 percent, and 25 percent, respectively.

It should be noted that failure to meet a reliability requirement does not necessarily result in a system's not being promoted to full-rate production. The DOT&E *2011 Annual Report* summarizes operational reliability and operational suitability results for 52 system evaluations that DOT&E provided to Congress for 2006 to 2011: a full 50 percent of the systems failed to meet their reliability threshold, and 30 percent of the systems were judged to be unsuitable. However, none of the 52 systems was cancelled.

DoD estimates that 41 percent of operational tests for acquisition category I (ACAT I) defense systems met reliability requirements between 1985 and 1990, while only 20 percent of such tests met reliability requirements between 1996 and 2000.[2] For Army systems alone, the Defense Science Board report on developmental test and evaluation (U.S. Department of Defense, 2008a) plotted estimated reliabilities in comparison with requirements for all operational tests of ACAT I Army systems between 1997 and 2006: only one-third of the systems met their reliability requirements. The Defense Science Board also found substantial declines in the percentage of Navy systems meeting their reliability requirements from 1999 to 2007.[3] These plots strongly indicate that there was an increasing problem in DoD regarding the ability to achieve reliability requirements in recent defense acquisition programs, especially for the higher-priced systems (those in which the Office of the Secretary of Defense is obligated to become involved) between 1996 and 2007.

At the workshop, Kendall stressed the importance of reliability engineering to address DoD's acknowledged reliability deficiency. He said that the department has not brought sufficient expertise in systems engineering to bear on defense acquisition for many years, and, as a result, many defense systems that have been recently deployed have not attained their anticipated level of reliability either in operational testing or when fielded. He said that this problem became very serious during the mid-1990s, when various DoD acquisition reform policies were instituted: they resulted in the elimination of sets of military standards; the relinquishment by the Office of the Secretary of Defense (OSD) of a role in overseeing quality control, systems engineering, reliability, and developmental testing; and associated severe cuts in staffing in service and program offices.

In a presentation to the International Test and Evaluation Association on January 15, 2009, Charles McQueary, then director of DOT&E, said that DOT&E needed to become more vigilant in improving the reliability of defense systems: for example, in 2008 two of six ACAT I systems in

[2]U.S. Department of Defense (2005, pp. 1-4).

[3]U.S. Department of Defense (2008a, pp. 3, 18)

operational testing were found not suitable.[4] This was a particularly serious issue because sustainment costs, which are largely driven by reliability, represent the largest fraction of system life-cycle costs. Also, as systems are developed to remain in service longer, the importance of sustainment costs only increases.

McQueary stressed that small investments in improving reliability could substantially reduce life-cycle costs. He provided two specific examples: two Seahawk helicopters, the HH-60H and the MH-60S. For one, an increased reliability of 2.4 hours mean time to failure would have saved $592.3 million in the estimated 20-year life-cycle costs; for the other, an increased reliability of 3.6 hours mean time to failure would have saved $107.2 million in the estimated 20-year life-cycle costs.[5] (A good analysis of the budgetary argument can be found in Long et al., 2007.)

Several years ago, the Defense Science Board issued a report with a number of findings and recommendations to address reliability deficiencies. It included the following key recommendation (U.S. Department of Defense, 2008a, pp. 23-24):

> The single most important step necessary to correct high suitability failure rates is to ensure programs are formulated to execute a viable systems engineering strategy from the beginning, including a robust RAM [reliability, availability, and maintainability] program, as an integral part of design and development. *No amount of testing will compensate for deficiencies in RAM program formulation* [emphasis added].

In other words, it is necessary to focus on system engineering techniques to design in as much reliability as possible at the initial stage of development. The result of inadequate initial design work is often late-stage adjustments of system design, and such redesigning of a system to address reliability deficiencies after a design is relatively fixed is more expensive than addressing reliability during the initial stages of system design. Moreover, late-stage design changes can often result in the introduction of other problems in a system's development.

The report of the Defense Science Board (U.S. Department of Defense, 2008a, p. 27) also contained the following finding:

---

[4]ACAT I—acquisition category I programs—are those for major defense acquisitions. They are defined by USD AT&L as requiring eventual expenditures for research, development, testing, and evaluation of more than $365 million (in fiscal 2000 constant dollars); requiring procurement expenditures of more than $2.19 billion (in fiscal 2000 constant dollars); or are designated as of high priority.

[5]See Chapter 3 for a discussion of mean time to failure.

> The aggregate lack of process guidance due to the elimination of specifications and standards, massive workforce reductions in acquisition and test personnel, acquisition process changes, as well as the high retirement rate of the most experienced technical and managerial personnel in government and industry has a major negative impact on DoD's ability to successfully execute increasingly complex acquisition programs.

Over the past 5 years, DoD has become more responsive to the failure of many defense systems to meet their reliability requirements during development. In response, the department has produced or modified a number of its guidances, handbooks, directives, and related documents to try to change existing practices. These documents support the use of more up-front reliability engineering, more comprehensive developmental testing focused on reliability growth, and greater use of reliability growth modeling for planning and other purposes.[6]

Two important recent documents are DTM-11-003 (whose improvements have been incorporated into the most recent version of DoDI 5000.02)[7] and ANSI/GEIA-STD-0009[8] (for details, see Appendix C). Although ANSI/GEIA-STD-0009 does not have an obligatory role in defense acquisition, it can be agreed upon by the acquisition program manager and the contractor as a standard to be used for the development of reliable defense systems.

We are generally supportive of the contents of both of these documents. They help to produce defense systems that have more reasonable reliability requirements and that are more likely to meet these requirements through design and development. However, these documents were designed to be relatively general, and they are not intended to provide details regarding specific techniques and tools or for engineering a system or component for high reliability. Nor do they mandate the methods or tools a developer would use to implement the process requirements. The tailoring will depend on a "customer's funding profile, developer's internal policies and procedures and negotiations between the customer and developer" (ANSI/GEIA-STD-0009, p. 2).[9] Proposals are to include a reliability program plan, a conceptual reliability model, an initial reliability flow-down of require-

---

[6]For a discussion of some of these documents, see Appendix C.

[7]A DTM (Directive-Type Memo) is a memorandum issued by the Secretary of Defense, Deputy Secretary of Defense, or OSD principal staff assistants that cannot be published in the DoD Directives System because of lack of time to meet the requirements for implementing policy documents.

[8]While not a DoD standard, ANSI/GEIA-STD-0009, "Reliability Program Standard for Systems Design, Development, and Manufacturing," was adopted for use by DoD in 2009.

[9]Available: http://www.techstreet.com/publishers/285174?sid=msn&utm_source=bing&utm_medium=cpc [August 2014].

ments, an initial system reliability assessment, candidate reliability trade studies, and a reliability requirements verification strategy.

But there is no indication of how these activities are to be carried out. For example, how should one produce the initial reliability assessment for a system when it only exists in diagrams? What should design for reliability entail, and how should it be carried out for different types of systems? How can one determine whether a test plan is adequate to take a system with a given initial reliability and improve that system's reliability to the required level through test-analyze-and-fix? How should reliability be tracked over time, with a relatively small number of developmental or operationally relevant test events? How does one know when a prototype for a system is ready for an operational test?

A handbook has been produced with one goal of providing more operational specificity,[10] but it understandably does not cover all the questions and possibilities. We believe that it would be worthwhile for an external group to assist in the provision of some additional specificity as to how some of these steps should be carried out.

Given the lengthy development time of ACAT I systems, the impact of the introduction of these new guidances and standards and memoranda, especially the changes to DODI 5000.02 (due to DTM 11-003) and ANSI-GEIA-STD-0009, will not be known for some time. However, we expect that adherence to these documents will have very positive effects on defense system reliability. We generally support the many recent changes by OSD. In this report, we offer analysis and recommendations that build on those changes, detailing the engineering and statistical issues that still need to be addressed.

KEY TERMS IN DEFENSE ACQUISITION

The assessment of defense systems is typically separated into two general operational assessments: the assessment of system effectiveness and the assessment of system suitability:

- *Operational effectiveness* is the "overall degree of mission accomplishment of a system when used by representative personnel in the environment planned or expected for operational employment of the system considering organization, doctrine, tactics, survivability or operational security, vulnerability, and threat." (U.S. Department of Defense, 2013a, p. 749).

[10]The handbook was produced by TechAmerica, *TechAmerica Engineering Bulletin Reliability Program Handbook, TA-HB-0009*. Available: http://www.techstreet.com/products/1855520 [August 2014].

- *Operational suitability* is "the degree in which a system satisfactorily places in field use, with consideration given to reliability, availability, compatibility, transportability, interoperability, wartime usage ranges, maintainability, safety, human factors, manpower, supportability, logistics supportability, documentation, environmental effects, and training requirements" (U.S. Department of Defense, 2013a, pp. 749-750).

Essentially, system effectiveness is whether a system can accomplish its intended missions when everything is fully functional, and system suitability is the extent to which, when needed, the system is fully functional.

Reliability is defined as "the ability of a system and its parts to perform their mission without failure, degradation, or demand on the support system under a prescribed set of conditions" (U.S. Department of Defense, 2012, p. 212). It can be measured in a number of different ways depending on the type of system (continuously operating or one-shot systems), whether a system is repairable or not, whether repairs return the system to "good as new," and how a system's reliability changes over time. For continuously operating systems that are not repairable, a common DoD metric is mean time to failure (see Chapter 3).

The evaluation of system suitability often involves two other components, availability and maintainability. *Availability* is "the degree to which an item is in an operable state and can be committed at the start of a mission when the mission is called for at an unknown (random) point in time" (U.S. Department of Defense, 2012, p. 214). *Maintainability* is "the ability of an item to be retained in, or restored to, a specified condition when maintenance is performed by personnel having specified skill levels, using prescribed procedures and resources, at each prescribed level of maintenance and repair" (U.S. Department of Defense, 2012, p. 215). Most of this report concentrates on development and assessment of system reliability, though some of what we discuss has implications for the assessment of system availability. The three components of suitability, reliability, availability, and maintainability, are sometimes referred to as RAM.

Ensuring system effectiveness does and should take precedence over concerns about system suitability. After all, if a system cannot carry out its intended mission even when it is fully functional, then certainly the degree to which the system is fully functional is much less important. However, this subordination of system suitability has been overdone. Until recently, DoD has focused the great majority of its design, testing, and evaluation efforts on system effectiveness, under the assumption that reliability (suitability) problems can be addressed either later in development or after initial fielding (see, e.g., U.S. Government Accountability Office, 2008).

THE STAGES OF DEFENSE ACQUISITION

The production of reliable defense systems begins, in a sense, with the setting of reliability requirements, which are expected to be (1) necessary for successful completion of the anticipated missions, (2) technically attainable, (3) testable, and (4) associated with reasonable life-cycle costs. After a contract is awarded, DoD has to devote adequate funds, oversight, and possibly additional time in development and testing to support and oversee both the application of reliability engineering techniques at the design stage and testing focused on reliability during contractor and government testing. These steps greatly increase the chances that the final system will satisfy its reliability requirements. The reliability engineering techniques that are currently used in industry to produce a system design consistent with a reliable system prior to reliability testing are referred to collectively as "design for reliability" (see Chapters 2 and 5). After the initial design stage, various types of testing are used to improve the initial design and to assess system reliability. A set of models are used throughout this development process to help oversee and guide the application of testing, and are commonly referred to as reliability growth models (see Chapter 4).

After the design stage, defense systems go through three phases of testing. "Contractor testing" is a catchall term for all testing that a contractor conducts during design and development, prior to delivery of prototypes to DoD. Contractor testing is initially component- and subsystem-level testing of various kinds. Some is in isolation, some is with representation of interfaces and interoperability, some is under laboratory conditions, some is under more realistic operating conditions, and some is under accelerated stresses and loads. Contractor testing also includes full-system testing after testing of components and subsystems: the final versions of these tests should attempt to emulate the structure of DoD operational testing so that the prototypes have a high probability of passing operational tests. There are obvious situations in which the degree to which the contractor's testing can approximate operational testing is limited, including some aircraft and ship testing.

Contractor testing, at least in the initial phase, ends with the delivery of prototypes to DoD for its own testing. The first phase of DoD testing is developmental testing, which is often initially focused on component- and subsystem-level testing; later, it focuses on full-system testing. There can be many respects in which developmental testing does not fully represent operational use. First, developmental testing is generally fully scripted, that is, the sequence of events and actions of the friendly and enemy forces, if represented, is generally known in advance by the system operators. Also, developmental testing does not often involve typical users as operators or typical maintainers. Furthermore, developmental testing often fails to fully represent the activities of enemy systems and countermeasures. However, in some situations, developmental testing can be more stressful than would be experienced in operational use, most notably in accelerated testing. The full process of government developmental testing is often conducted over a number of years.

After developmental testing, DoD carries out operational testing. Initial operational testing, which is generally a relatively short test of only a few months duration, is full-system testing under as realistic a set of operational conditions as can be produced given safety, noise, environmental, and related constraints. Operational testing is much less scripted than developmental testing and uses typical maintainers and operators. It is used as a means of determining which systems are ready to be promoted to full-rate production, i.e., deployed. Toward this end, measurements of key performance parameters collected during operational testing are compared with the associated requirements for effectiveness and suitability, with those systems successfully meeting their requirements being promoted to the field.

Ideally, developmental testing would have identified the great majority of causes of reliability deficiency prior to operational testing, so that any needed design changes would have been recognized prior to full specification of the system design. Such recognition would have resulted in design changes that would be less expensive to implement at that stage than later. Furthermore, because operational testing is not well designed to discover many reliability deficiencies because of its fairly limited time frame, it should not be depended on to capture a large number of such problems.

Moreover, because developmental testing often does not stress a system with the full spectrum of operational stresses, it often fails to discover many design deficiencies, some of which then surface during operational testing. This failure could also be due, at times, to changes in the data collection techniques and estimation methodology in the testing. There is no requirement for consistent failure definitions and scoring criteria across developmental and operational testing. In fact, as Paul Ellner[11] described at the panel's workshop, for a substantial percentage of defense systems, reliability as assessed in operational testing is substantially lower than reliability of the same system as assessed in developmental testing. This difference is often not explicitly accounted for in assessing which systems are on track to meet their requirements: this lack of recognition of the difference may in turn account for the failure of many systems to meet their reliability requirements in operational testing after being judged as making good progress toward the reliability requirement in developmental testing and evaluation.

[11]Presentation to the panel at its September 2011 workshop.

A HARD PROBLEM, BUT PROGRESS IS POSSIBLE

ACAT I defense systems, the systems that provide the focus of this report, are complicated.[12] They can generally be represented as systems of systems, involving multiple hardware and software subsystems, each of which may comprise many components. The hardware subsystems sometimes involve complicated electronics, and the software subsystems often involve millions of lines of code, all with interfaces that need to support the integration and interoperability of these components, and all at times operating under very stressful conditions.

While defense systems are growing increasingly complex, producing reliable systems is not an insurmountable challenge, and much progress can be made through the use of best industrial practices.

There are, however, important differences between defense acquisition and industrial system development (see Chapter 2). For instance, defense acquisition involves a number of "actors" with somewhat different incentives, including the contractor, the program manager, testers, and users, which can affect the degree of collaboration between DoD and the contractor. Furthermore, DoD assumes the great majority of risk of development, which is handled in the private sector through the use of warranties and other incentives and penalties. Acknowledgment of these distinctions has implications as to when and how to best apply design for reliability, reliability testing, and formal reliability growth modeling.

In this report, we examine the applicability of industrial practices to DoD, we assess the appropriateness of recent reliability enhancement initiatives undertaken by DoD, and we recommend further modifications to current DoD acquisition processes.

As noted, in addition to the use of existing design for reliability and reliability testing techniques, we were asked to review the current role of formal reliability growth models. These models are used to plan reliability testing budgets and schedules, track progress toward attaining requirements, and predict when various reliability levels will be attained. Reliability growth is a result of design changes or corrective actions resulting, respectively, from engineering analysis or from correction of a defect that becomes apparent from testing. Often included in reliability growth modeling are fix effectiveness factors, which estimate the degree to which design changes are fully successful in eliminating a reliability failure mode. Formal reliability growth models are strongly dependent on often unvalidated assumptions about the linkage between time on test and the discovery of reliability defects and failure modes. The impact of relying on these unvali-

---

[12]Though ACAT I systems are a focus, the findings and recommendations contained here are very generally applicable.

dated assumptions can result in poor inferences through use of the predicted values and other model output. Therefore, the panel was asked to examine the proper use of these models in the three areas of application listed above.

The goal of this report is to provide advice as to the engineering, testing, and management practices that can improve defense system reliability by promoting better initial designs and enhancing prospects for reliability growth through testing as the system advances through development. We include consideration of the role of formal reliability growth modeling in assisting in the application of testing for reliability growth.

There is a wide variety of defense systems and there are many different approaches toward the development of these systems in several respects. Clearly, effective methods differ for different kinds of systems, and therefore, there can be no recommended practices for general use. Furthermore, while there is general agreement that there have been problems over the past two decades with defense system reliability, even during this period there were defense systems that used state-of-the-art reliability design and testing methods, which resulted in defense systems that met and even exceeded their reliability requirements. The problem is not that the methods that we are describing are foreign to defense acquisition: rather, they have not been consistently used, or have been planned for use but then cut for budget and schedule considerations.

REPORT STRUCTURE

The remainder of this report is comprised of nine chapters and five appendixes. Chapter 2 reports on the panel's workshop, which focused on reliability practices in the commercial sector and their applicability to defense acquisition. Chapter 3 discusses different reliability metrics that are appropriate for different types of defense systems. Chapter 4 discusses the appropriate methods and uses for formal reliability growth modeling. Chapter 5 covers the tools and techniques of design for reliability. Chapter 6 documents the tools and techniques of reliability growth testing. Chapter 7 discusses the design and evaluation of reliability growth testing relevant to developmental testing. Chapter 8 details the design and evaluation of reliability growth testing relevant to operational testing. Chapter 9 covers software reliability methods. Chapter 10 presents the panel's recommendations. Appendix A lists the recommendations of previous reports of the Commitee on National Statsitics that are relevant to this one. Appendix B provides the agenda for the panel's workshop. Appendix C describes recent changes in DoD formal documents in support of reliability growth. Appendix D provides a critique of MIL HDBK 217, a defense handbook that provides information on the reliability of electronic components. Finally, Appendix E provides biographical sketches of the panel members and staff.

2

Defense and Commercial System Development: A Comparison

The role of the U.S. Department of Defense (DoD) in providing oversight and management of the acquisition process was emphasized by our sponsor as an issue that might benefit from the panel's review. To help us understand those issues as broadly as possible, we undertook a brief review of system development for commercial products, in particular, of those companies recognized as producing products with high reliability. This topic was a feature of the panel's workshop, and it informed our work.

Throughout this report we often make comparisons, both implicit and explicit, between system development as practiced by companies that are recognized as producing highly reliable systems and current practice in defense acquisition. Such comparisons are useful, but it is important to keep in mind that system development in defense acquisition has important differences that distinguish it from commercial system development.

The first section below discusses key differences between commercial and defense acquisition. The second section discusses the use of an incentive system for defense acquisition with regard to reliability. The final section of the chapter presents a view of best practices by Tom Wissink of Lockheed Martin and Lou Gullo of Raytheon, who have extensive experience in developing reliable defense systems.

THREE KEY DIFFERENCES

The first difference between commercial and defense acquisition is the sheer size and complexity of defense systems. In developing, manufacturing, and fielding a complicated defense system (e.g., an aircraft or land vehicle), one is dealing with many subsystems, each of substantial complexity: this characteristic, by itself, poses enormous management and technological challenges. A ship, for instance, might have a single program manager but more than 100 acquisition resource managers. Or a defense system may be a single element in an extensive architecture (e.g., command, control, and communications [C3] network), with separate sets of interface requirements evolving as the architecture components progress through their individual stages of acquisition and deployment. New systems may also have to interface with legacy systems. Furthermore, defense systems often strive to incorporate emerging technologies. Although nondefense systems can also be extremely complicated and use new technologies, they often are more incremental in their development, and therefore tend to be more closely related to their predecessors than are defense systems.

Second, there are significant differences in program management between commercial and defense system development. Commercial system development generally adheres to a single perspective, which is embodied in a project manager with a clear profit motive and direct control of the establishment of requirements, system design and development, and system testing. In contrast, defense system development has a number of relatively independent "agents," including the system development contractor; a DoD program manager; DoD testers; and the Office of the Secretary of Defense, which has oversight responsibilities. There is also the military user, who has a different relationship with the contractor than a customer purchasing a commercial product from a manufacturer (see below).

These different groups have varying perspectives and somewhat different incentives. In particular, there is sometimes only limited information sharing between contractors and the various DoD agents. This lack of communication is a serious constraint when considering system reliability, because it prevents DoD from providing more comprehensive oversight of a system's readiness for developmental and operational testing with respect to its reliability performance. It also limits DoD's ability to target its testing program to aspects of system design that are contributing to deficient system reliability. Once prototypes have been provided to DoD for developmental testing, let alone for operational testing, the system design is relatively set, and cost-efficient opportunities for enhancing reliability may have been missed.

A third difference between defense and commercial system development concerns risk. In the commercial world, the manufacturer carries most of the risks that would result from developing a system with poor reliability performance. Such risks include low sales, increased warranty expenses, the loss of customer goodwill for products with poor reliability, and increased life-cycle costs for systems that are maintained by the manufacturer (e.g., locomotives and aircraft engines). Consequently, commercial manufacturers have a strong motivation to use reliability engineering methods that help to ensure that the system will meet its reliability goals by the time it reaches the customer.

For defense systems, however, the government and the customers (i.e., the military services that will use the system) generally assume most of the risk created by poor system reliability. Therefore, the system developers do not have a strong incentive to address reliability goals early in the acquisition process, especially when it can be difficult to quantify the possibly substantial downstream benefits of (seemingly) costly up-front reliability improvement efforts.

This third issue is extremely important: if developers shared some of the risk, then they would be much more likely to focus on reliability early in development, even without significant DoD oversight. Risk sharing is one component of the larger question of establishing a system of rewards and penalties—either during development or after delivery—for achieving (or exceeding) target levels or final requirements for reliability or for failing to do so.

Although DoD has at times used incentives to reward contractors for exceeding requirements, the panel is unaware of any attempt to institute a system of warranties for a defense system that provides for contractor payments for failure to meet reliability requirements.[1] And in considering incentives, we are unaware of any studies of whether offering such payments has succeeded in motivating developers to devote greater priority to meeting reliability requirements. Most relevant to this report is the idea that rewards could be applied at intermediate points during development, with rewards for systems that are assessed to be ahead of intermediate reliability goals, or penalties for systems that are assessed to be substantially behind such goals. (See discussion in Chapter 7.)

ISSUES IN AN INCENTIVE SYSTEM FOR DEFENSE ACQUISITION

In considering an incentive system, we note some complications that need to be kept in mind in its development. First, such a system must be based on the recognition that some development problems are intrinsically harder than others by requiring a nontrivial degree of technology development. Having a penalty for not delivering a prototype on time or not meeting a reliability requirement may dissuade quality developers from bidding on such development contracts because of the perceived high risk. The best contractors may be the ones that know how difficult it will be to meet a requirement and might therefore not offer a proposal for a difficult acquisition program because of a concern about incurring penalties because

[1]For a useful consideration of statistical issues involved with product warranties, see Blischke and Murthy (2000).

of something that could not be anticipated. Therefore, it would be useful, to the extent possible, to link any incentive payments, either positive or negative, to the intrinsic challenge of a given system.

Looking at a later stage in the acquisition process, providing incentives (or penalties) for performance after delivery of prototypes, rather than during intermediate stages of development, would have the advantage of being able to use more directly relevant information, in that the the reliability level achieved by a system could be directly assessed through DoD developmental and operational testing. But for earlier, intermediate reliability goals, the assessment is carried out by the contractor, so an incentive system may lead to various attempts to affect ("game") the assessment. For instance, the test environments and the stress levels for those environments might be selected to avoid ones that are particularly problematic for a given system. Moreover, whether a test event should or should not be counted as a failure or whether a test event itself is considered "countable" can sometimes rely on judgment and interpretation. It is also important to note that testing for intermediate reliability goals is not likely to have large sample sizes, and such estimates are therefore likely to have substantial uncertainty. As a result and depending on the decision rule used, there are likely to be either substantial consumer or producer risks in such decisions concerning incentive payments.

Finally, offering incentives regarding delivery schedules raises another concern: potential tradeoffs between early delivery and reliability achievement. Because life-cycle costs and overall system performance can be sensitive to system reliability, compromising on reliability performance to meet an inflexible deadline is often a bad bargain.

A PERSPECTIVE ON COMMERCIAL BEST PRACTICES

For the panel's September 2011 workshop, we asked Tom Wissink of Lockheed Martin and Lou Gullo of Raytheon to discuss best practices that are currently used to develop reliable systems and how DoD could promote and support such actions in its requests for proposals and contracts. We asked them to include in their comments the use of design-for-reliability methods, reliability testing, and use of reliability growth models. We also asked them to comment on the impact of recent changes in procedures brought about by the adoption of ANSI/GEIA-STD-0009 and DTM 11-003 and to provide suggestions on how best to move forward to help produce more reliable defense systems in the future.[2] To answer our questions, Wissink and Gullo not only relied on their own experiences, but also talked with engineers at their companies. They presented their answers both in

[2]Please see footnotes 4 and 5 in Chapter 1 on the role of these two documents.

writing and orally at the workshop. Overall, Wissink and Gullo said, design for reliability requires

- identification and mitigation of design weaknesses,
- detection and resolution of mission critical failures, and
- characterization of design margins and continuous design improvements to decrease failures in the field and reduce total cost of ownership over the system or product life cycle.

Regarding the communication of the need for design for reliability in proposals, they said that every acquisition contract should specify (1) the system reliability requirement as a key performance parameter, (2) what reliability engineering activities should be performed during system development to achieve the requirement and (3) the means to verify the reliability requirement. "Reliability growth management as part of design for reliability represented in proposals is a way to specifically require actions to improve reliability over time with reliability assessment of standard reliability metrics over the system/product life cycle, starting early in the development phase," Wissink and Gullo wrote.

Winnick and Gullo suggested that the DoD acquisition requests for proposals should ask for a written reliability growth management plan and a reliability growth test plan as part of the integrated test planning. There should also be a "reliability growth incentive award scale and incentive fee scheduled during intervals in the development cycle so that the contractor is rewarded for favorable reliability growth that exceeds customer expectations." Reliability growth management planning entails the development of reliability growth curves for the system, major subsystems, products, and assemblies, along with the plan for achieving specified reliability values.

Reliability growth management "includes reliability assessments and a test plan that contains various types of testing (e.g., accelerated life tests, highly accelerated life tests), adequate test time in the program schedule, and test samples to demonstrate increasing reliability with increasing confidence over time." Furthermore, they wrote, reliability growth management provides "a means for tracking reliability growth from system level to assembly or configuration item level, and monitoring progress. . . ." It also includes "intervals in the development phase to allow for implementation of design change corrective actions to positively affect reliability with each subsequent design modification."

Winnick and Gullo expanded on what DoD should require in a request for proposal (RFP). They said that the RFP should require that the different support contractors for each major subsystem provide a reliability growth profile[3] to demonstrate the anticipated reliability growth over time. This curve should be a member of a set of models that are approved by DoD for such use. The software used to implement this should provide standard reporting outputs to program management to verify and validate the reliability growth curve for the particular program.

The outputs, they said, should include plotting the reliability growth curves over time (for each major subsystem) to provide the following information:

- the expected starting point (initial reliability), with the context in a separate document that supports the data sources and the rationale for its selection;
- the number of tests planned during the development program to be used to verify that starting point;
- the expected reliability growth profile with the context in a separate document that supports the data sources and the rationale for the selection of the points on the graph;
- the number of tests needed to produce that profile, the schedule for these tests, and the schedule for implementing design change corrective actions for the failures that are expected to occur, resulting in design reliability improvements; and
- a risk assessment for the starting point, reliability growth profile, and number of tests necessary to meet the required reliability levels on the growth curve.

They also noted that every DoD acquisition RFP and test and evaluation master plan or systems engineering plan "should require contractors to provide a reliability growth management plan and a reliability growth test plan as part of the integrated test plan and reliability growth profile."

The presentation by Winnick and Gullo and the workshop discussion on this topic provided the panel with a better understanding of what contractors would be willing to provide in support of the production of reliable systems. We note, however, several reservations about their conclusions.

There is a great deal of variability in the complexity of DoD systems and the constitution of their respective "major subsystems." We are not convinced that a formal mathematical reliability growth curve should be developed for every major subsystem of every DoD system—although some sound plan for growing and demonstrating reliability is appropriate for each major subsystem. The approval of the reliability growth tools that are implemented for a specific subject system and its major subsystems is tacit

---

[3]A reliability growth profile is a representation of future reliability growth as a function over time: see Chapter 4.

in the DoD oversight of acquisition processes, which includes the review and approval of essential test and evaluation documents. Furthermore, we do not envision the construction of a master list of DoD-approved reliability growth tools. As discussed throughout the report, the viability and appropriateness of such tools will need to be case specific.

3

Reliability Metrics

In the context of the U.S. Department of Defense (DoD) acquisition system, reliability metrics are summary statistics that are used to represent the degree to which a defense system's reliability as demonstrated in a test is consistent with successful application across the likely scenarios of use. Different metrics are used in conjunction with continuously operating systems (such as tanks, submarines, aircraft), which are classified as either repairable or nonrepairable, and with "one-shot" systems (such as rockets, missiles, bombs). Reliability metrics are calculated from data generated by test programs.

This chapter discusses "reliability metrics," such as the estimated mean time between failures for a continuously operating system. We consider repairable and nonrepairable systems, continuous and one-shot systems, and hybrids.

A system's requirements in a request for proposal (RFP) will be written in terms of such metrics, and these metrics will be used to evaluate a system's progress through development. Tracking a reliability metric over time, as a system design is modified and improved, leads to the topic of reliability growth models, which are the subject of the next chapter.

CONTINUOUSLY OPERATING REPAIRABLE SYSTEMS

In developmental and operational testing, continuously operating systems that are repairable perform their functions as required until interrupted by a system failure that warrants repair or replacement (ordinarily at the subsystem or component level). For measuring and assessing operational reliability, the primary focus for "failures" is generally restricted to operationally critical failure modes, which include operational mission failure, critical failure, and system abort. Test results and requirements are often expressed accordingly—as the mean time between operational mission failures (MTBOMF), the mean time between critical failures (MTBCF), and the mean time between system abort (MTBSA)—or as the probability of successfully completing a prescribed operational mission of a given time duration without experiencing a major failure.[1,2]

Standard DoD reliability analyses normally entail three analytical assumptions:

1. Restoration activities return a failed test article to a state that is "as good as new." That is, the time to first failure (from the beginning of the test[3]) and the subsequent times between failures for the subject test article all are taken to be statistically independent observations governed by a single probabilistic distribution.
2. The same time-to-failure distribution (or failure probability for one-shot systems) applies to each test article over replications.
3. The common time-to-failure distribution (or failure probability for one-shot systems) is exponential with failure rate parameter λ (alternatively parameterized in terms of a mean time to failure parameter, $\theta = 1/\lambda$).

There are two advantages to invoking this nominal set of assumptions: it simplifies statistical analyses, and it facilitates the interpretability of results. Analyses are then the examination of the number of failure times and censored times (from the time of the last failure for a test article to the end of testing time for that article) that are observed, assuming a single underlying exponential distribution. A mathematically equivalent formulation is that the total number of observed failures in the total time on test (across all test articles), T, is governed by a Poisson distribution with expected value equal to λT (or T/θ). The DoD primer on reliability, availability, and maintainability (RAM) (U.S. Department of Defense, 1982), as well as numerous textbooks on reliability, address this nominal situation (within the framework of a homogeneous Poisson process) and provide straightforward estimation, confidence bounds, and test duration planning

[1]Lower levels of failures should not necessarily be ignored by logistics planners, especially if they will lead to substantial long-term supportability costs.

[2]Another important metric, but outside of the scope of this report, is operational availability—the long-term proportion of time that a system is operationally capable of performing an assigned mission. Estimating availability requires estimating system down times due to planned and unplanned maintenance activities.

[3]Test articles may undergo pretest inspections and maintenance actions.

methodologies. In practice, the customary estimate of a system's mean time between failures is simply calculated to be the total time on test, *T*, divided by the total number of observed failures (across all test articles).[4] It is readily comprehensible and directly comparable to the value of reliability that had been projected for the test event or required to be demonstrated by the test event.

Although the above assumptions support analytical tractability and are routinely undertaken in DoD assessments of the reliability demonstrated in an individual test, alternative assumptions merit consideration and exploration for their viability and utility. Rather than the assumption of a return to a state "as good as new," a physically more defensible assumption in many instances (e.g., in a system with many parts) might be that a repair or replacement of a single failed part only minimally affects the system state (relative to what it was before the failure) and the system is thus restored to a state approximately "as bad as old." This perspective would accommodate more complex phenomena in which the system failure rate may not be constant over time (e.g., monotonically increasing, which corresponds to aging articles that tend to experience more failures as operating time accumulates). Flexible statistical models and analysis approaches suitable for these more general circumstances, both parametric and nonparametric, are widely available (e.g., Rigdon and Basu, 2000; Nelson, 2003). Sample size demands for precise estimation, however, may exceed what typical individual developmental or operational tests afford. For example, the total hours on test available for a single test article often can be quite limited—spanning only a few lifetimes (measured in terms of the prescribed reliability requirement) and sometimes even less than one lifetime. An additional issue relates to the interpretability of models that portray nonconstant failure intensities: In particular, what sort of a summary estimate for a completed developmental or operational test event should be reported for comparison to a simply specified mean time between failure prediction or requirement (that did not contemplate time-variant intensities)?

Sample size limitations likewise may hinder examinations of heterogeneity for individual or collective groupings of test articles. When data are ample, statistical tests are available for checking a number of potential hypotheses of interest, such as no variability across select subgroups (e.g., from different manufacturing processes), no outliers that may be considered for deletion from formal scoring and assessment (e.g., possibly attributable to test-specific artificialities), and the like. Caution needs to be taken, however, to recognize potential sensitivities to inherent assumptions (e.g., such

[4]Under the three analytic assumptions above, the mean time between failures is synonymous with the mean time to failure or the mean time to first failure, other commonly used terms.

as assumptions 1 and 3 above) attendant to the application of any specific methodology.

Although there often are plausible motivations for asserting that an exponential time-to-failure distribution is appropriate (e.g., for electronics or for "memoryless" systems in circumstances for which aging is not a major consideration), there is no scientific basis to exclude the possibility of other distributional forms. For example, the more general two-parameter Weibull distribution (which includes the exponential as a special case) is frequently used in industrial engineering. Observed failure times and available statistical goodness-of-fit procedures can guide reliability analyses to settle on a particular distribution that reasonably represents the recorded data from a given test. The plausibility of the "as good as new" assumption warrants scrutiny when repeat failures (recorded on an individual test article) are incorporated into the analyses.

Distinct estimation and confidence interval methods are associated with different choices for the time-to-failure distribution. The mathematical form of the distribution function provides a direct link between parameter estimates (e.g., the mean time between failures for the exponential distribution) and the probability of the system performing without a major failure over a prescribed time period (e.g., mission reliability). For a given set of failure data, different specifications of the time-to-failure distribution can lead to different estimates for the mean time between failures and generally will lead to distinct estimates for mission reliability. For the one-parameter exponential distribution, there is a one-to-one correspondence between the mean time between failures and mission reliability. This is not the case for other distributions.

Implicit in assumption 3 above is that the environment and operating conditions remain constant for the test article each time it is repaired and returned to service. Unless statistical extrapolation methods are applied, reliability estimates generated from a single test's observed failure data should be interpreted as representative solely of the circumstances of that test.[5] The possible effects of influential factors (e.g., characterizing the execution of the testing or description of the past usage or maintenance and storage profiles) on system reliability can be portrayed in regression models or hierarchical model structures. For instance, with sufficient test data, one could assess whether changes in storage conditions had an impact on system reliability. In general, adequate sample sizes would be needed to support parameter estimation for these more sophisticated representations of system reliability.

[5]The issue of the relevance of the testing conditions to operational scenarios is considered in the next chapter.

CONTINUOUSLY OPERATING NONREPAIRABLE SYSTEMS

Continuously operating nonrepairable systems (e.g., batteries, remote sensors) function until a failure occurs or until there is some signal or warning that life-ending failure is imminent: when that occurs, the system is swapped out. Each system experiences at most one failure (by definition—it cannot be restored and brought back into service after having failed). Some systems that routinely are subjected to minor maintenance can be viewed as nonrepairable with respect to catastrophic failure modes (e.g., a jet engine).

For these systems, a relevant reliability metric is mean time to failure. From an experimental perspective, nonrepairable systems can be tested until they fail or can be tested under censoring schemes that do not require all articles to reach their points of failure. These data provide an estimate of expected operational lifetimes and their variability. In addition, analytical models can be developed to relate expected remaining life to concomitant data, such as recorded information on past environmental or usage history, measures of accumulated damage, or other predictive indicators obtained from sensors on the systems.

Nonrepairable systems are common in many commercial settings, but they are rare in DoD acqusition category I testing. The role of prognostic-based reliability predictions to assess various forms of degradation in reducing defense system life-cycle costs, however, continues to gain prominence (Pecht and Gu, 2009; Collins and Huzurbazar, 2012). System program managers are instructed (U.S. Department of Defense, 2004, p. 4) to optimize operational readiness with "diagnostics, prognostics, and health management techniques in embedded and off-equipment applications when feasible and cost-effective."

ONE-SHOT SYSTEMS

Testing of one-shot (or "go/no go") systems in a given developmental or operational testing event involves a number of individual trials (e.g., separate launches of different missiles) with the observed performance in any trial being characterized as either a "success" or "failure." Assumption 1 above generally is not germane because a test article is required to function only once.[6] But assumptions 2 and 3 and the discussion above are, for the most part, very relevant. One exception is that the distribution of interest governing a single test result generally is modeled with a one-parameter Bernoulli distribution.

[6]There may be exceptions. For example, it is possible that a failure to launch for a rocket in a trial could be traced to an obvious fixable wiring problem and the rocket subsequently reintroduced into the testing program.

One associated reliability metric is the estimated probability of success. The estimate can be a "best estimate" or it can be a "demonstrated" reliability, for which the metric is a specified statistical lower confidence limit on the probability of success. Because reliability can depend on environmental conditions, the metrics of estimated reliability at specified conditions may be required, rather than a single reliability metric. It can be defined for individual prescribed sets of conditions that establish operational profiles and scenarios or for a predetermined collection of conditions. Estimates of system reliability derive from observed proportions of "successful" trials—either in total for a test event or specific to particular combinations of test factors (e.g., using logistic regression models). The estimates need to be interpreted to pertain to the specific circumstances of that testing. Given sufficient and adequate data, statistical modeling could support extrapolations and interpolations to other combinations of variables.

HYBRID MODELS

Hybrid models of system reliability, embodying both time-to-failure and success-failure aspects, also may be suitable for some testing circumstances. Imagine, for instance, a number of cruise missiles (without active warheads), which are dedicated to testing in-flight reliability, being repeatedly captive-carried by aircraft for extended periods to simulate the in-flight portions of an operational cruise missile mission. Observed system failures logically could be examined from the perspective of mean time between failures, facilitating the construction of estimates of in-flight reliability that correspond to distinct operational mission scenarios that span a wide spectrum of launch-to-target ranges. To obtain measures of overall reliability, these estimates could be augmented by results from separate one-shot tests, in the same developmental or operational testing event, that focus on the probabilities of successful performance for other non-in-flight elements of cruise missile performance (e.g., launch, target recognition, warhead activation, and warhead detonation).

In this example, the mode of testing (involving repairable test articles) does not match the tactical use of the operational system (single launch, with no retrieval). From an operational perspective, the critical in-flight reliability metric for cruise missiles could be taken to be mean time to failure—which can be conceptually different from mean time between failures when assumption 1 (above) does not hold.

ASSESSMENT OF CURRENT DoD PRACTICES

The process of deriving formal system reliability requirements and intermediate reliability goals to be attained at various intermediate test program points is undertaken well before any system testing begins. Consequently, the simple mean time between failures and probability-of-success metrics traditionally prescribed in DoD acquisition test and evaluation contexts are reasonable. For a given developmental or operational test event, knowledge of the nature of the reliability data and of the particulars of the testing circumstances, including composition and characteristics of test articles, are available. This information can support the development of alternative models and associated specific forms of reliability metrics that are potentially useful for describing reliability performance demonstrated in testing and projected to operational missions.

Very often, the standard mean time between failures and probability-of-success metrics will be appropriate for describing system-level reliability given the confines of the available test data. Such a determination should not be cavalierly invoked, however, without due consideration of more advanced plausible formulations—especially if those formulations might yield information that will support reliability or logistics supportability improvement initiatives, motivate design enhancements for follow-on testing, or substantively inform acquisition decisions. The more sophisticated methodological approaches based on more elaborate distributions with parameters linked to storage, transportation, type of mission, and environment of use may be particularly attractive after a system is fielded (for some classes of fielded systems), when the amount and composition of reliability data may change substantially given what is available from developmental and operational testing.

Several points that have been noted warrant emphasis. For any system, whether in the midst of a developmental program or after deployment, there is no such thing as a single true mean time between failures or actual mean time to first failure (Krasich, 2009). System reliability is a function of the conditions, stresses, and operating profiles encountered by the system during a given period of testing or operations, and these can and do vary over time. System reliability likewise is influenced by the composition of the systems themselves (i.e., test articles or specific deployed articles that are monitored), which may include diverse designs, manufacturing processes, past and current usage, and maintenance profiles. Estimates of the mean time between failures, mean time to first failure, or other metric for system reliability needs to be interpreted accordingly.

Operational reliability is defined in terms of one or more operational mission profiles expected to be encountered after a defense system has attained full-rate production status and is deployed. Ideally, system-level developmental and operational testing would mimic or plausibly capture the key attributes of these operational circumstances, particularly for operational testing and the later stages of developmental testing. It is important to understand, however, that there are limitations to the extent to which operational realism is or can be reflected in those testing events. Moreover, efficient developmental testing strategies, especially when a system's functional capabilities emerge incrementally over time, may not readily lend themselves to complete examination of system operational reliability, especially in the early stages of developmental testing. Again, as appropriate, distinctions should be drawn between estimates of system reliability and estimates of system operational reliability.

4

Reliability Growth Models

This chapter responds to a specific item in the charge to the panel, to consider reliability growth models in U.S. Department of Defense (DoD) acquisitions. Reliability growth models are models that are used to estimate or predict the improvement of system reliability as a function of the amount of system testing that is carried out. Such models are used in three ways: (1) to help construct test plans for defense systems early in development, (2) to assess the currently achieved system reliability, and (3) to assess whether a defense system in development is on track to meeting its reliability requirements prior to deployment. This chapter examines the form that these models have taken and some of their strengths and weaknesses for the three kinds of applications. We begin with an overview of the concept. We then look at the hardware growth models commonly used, their applications, and the implications for DoD use. Reliability growth models for software are covered in Chapter 9.

CONCEPTS AND EXAMPLES

The traditional DoD process for achieving reliability growth during development is known as test, analyze, and fix—TAAF. It includes system-level developmental test and posttest assessment of observed failures to determine their root causes. This assessment is followed by an analysis to identify potential reliability enhancements (e.g., hardware, software, manufacturing processes, maintenance procedures, or operations) and incorporation of specific design upgrades. The next step is retesting to verify that failure modes have been removed or mitigated and checking to see that no new failure modes have been introduced. During developmental testing, failed systems typically are restored (by repair or replacement of parts) and returned to testing. Repeated TAAF cycles during developmental test can "grow" system reliability over time.[1]

A typical profile for reliability growth is shown in Figure 4-1. Part a of the figure depicts real growth: growth occurs in incremental step increases, with larger gains occurring in the earlier tests—because failure modes with higher failure rates (or failure probabilities) are more likely to occur and contribute more to reliability growth when fixed than their counterparts with smaller failure rates (or probabilities of failure). For any specific system undergoing TAAF, the exact pattern and extent of the actual growth in reliability is random, because the occurrences of individual failure modes (which lead to reliability gains) are random events.

Part b in Figure 4-1 shows how reliability results from different TAAF phases (e.g., developmental test events) are connected. The primary objective is to use all of the available data to strengthen the statistical precision associated with an estimate of system reliability[2] (attained at the completion of testing of the last conducted event) and narrowing confidence intervals relative to what could be reported by using results solely from that last test event. The modeling also can smooth out point estimates of system reliability by accounting for inherent randomness in test observations.

There are three primary uses of the methodologies: to facilitate the planning of developmental testing programs (i.e., a series of TAAF phases); to track demonstrated reliability performance over the course of the testing; and to project anticipated reliability beyond observed testing.

Reliability growth modeling began with empirical observations by Duane (1964) on developmental testing programs for relatively complex aircraft accessories. For the systems he was tracking, on a log-log scale, the cumulative number of failures, $N(T)$, tended to increase linearly with the cumulative test time, T.[3] Since then, many reliability growth models have been developed (e.g., see the expository surveys in Fries and Sen, 1996; U.S. Department of Defense, 2011b).

Continuous time-on-test reliability growth models can be classified into two general categories. The first builds on a probabilistic framework for the

---

[1]The concept of reliability growth can be more broadly interpreted to encompass reliability improvements made to an initial system design *before* any physical testing is conducted, that is, in the design phase, based on analytical evaluations (Walls et al., 2005). Such a perspective can be useful for systems that are not amenable to operational testing (e.g., satellites).

[2]This estimate is made at the completion of the last test. It generally does not consider any reliability design improvements that may be implemented after the last event is completed and observed failure modes have been analyzed.

[3]This form of "Duane's Postulate," or "learning curve property," is equivalent to the *average* cumulative number of failures (i.e., $N(T)/T$) and is roughly linear in T on a log-log scale.

A

B

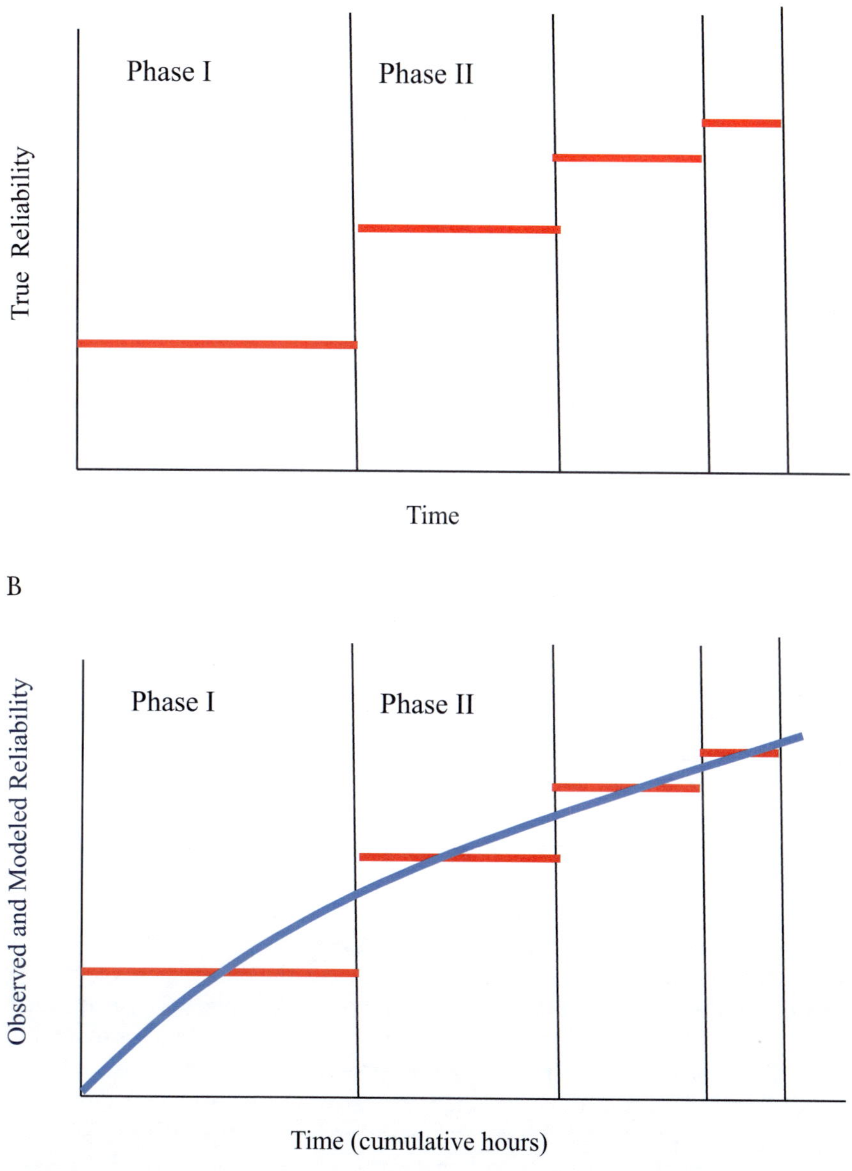

FIGURE 4-1 Illustrations of reliability growth using the TAAF (test, analyze, and fix) process.

cumulative count of failures over time. These are models of the underlying failure process. The second category directly imposes a mathematical structure on the time between successive failures (at the system level or for individual failure mode mechanisms), essentially modeling the dynamics of mean time between failures over time.[4] Within each category, a number of different approaches to this modeling have been taken. Other techniques have been adapted to the reliability growth domain from biostatistics, engineering, and other disciplines. Similar categorizations describe families of discrete reliability growth models (see, e.g., Fries and Sen, 1996).

Reliability growth models generally assume that the sole change between successive developmental testing events is the system reliability design enhancements introduced between the events. This assumption constrains their applicability because it specifically excludes the integration of reliability data obtained from substantially different testing circumstances (within a test or across test events). For example, laboratory-based testing in early developmental testing can yield mean-time-between-failure estimates that are considerably higher than the estimates from a subsequent field test. Similarly, the fact that successive developmental tests can occur in substantially different test environments can affect the assumption of reliability growth. For example, suppose a system is first tested at low temperatures and some failure modes are found and fixed. If the next test is at high temperatures, then the reliability may decline, even though the system had fewer failure modes due to the design improvements. Because most systems are intended for a variety of environments, one could argue that there should be separate reliability growth curves specific to each environment. This idea may be somewhat extreme, but it is critical to keep in mind that reliability growth is specific to the conditions of use.

Another characteristic shared by the great majority of reliability growth models is that any specific application imposes a common analysis treatment of the failure data across the entire testing program. Thus, there is a reduction in analytical flexibility for representing the results in individual developmental testing events. In addition, nearly all reliability growth models lack closed-form expressions for statistical confidence intervals. Asymptotic results have been derived for some models and conceptually are obtainable from likelihood function specifications—provided that proper care is taken to account for the non-independent structure of the failure event data. The availability of parametric bootstrap methods has the potential to support statistical inference across broad categories of reliability growth models, but to date the application of this tool has been limited.

[4]A model within one category necessarily generates a unique model from the other category. The physical interpretation that drives the modeling, however, does not translate readily from one type to another.

In DoD acquisition, a small number of reliability growth models dominate (see next section). But across applications, no particular reliability growth model is "best" for all potential testing and data circumstances.

The derivations of common reliability growth models are predominantly hardware-centric. In practice, however, their scope ordinarily encompasses software performance by using failure scoring rules that count all failures, whether traceable to hardware or to software failure modes, under a broad definition of "system" failure. However, the probabilistic underpinnings of software failure modes are quite different from those for hardware failure modes.[5] Nevertheless, the resultant forms of software reliability growth may serve to fit reliability data from general developmental test settings.

Given that the reliability of a complex system is in fact a multidimensional outcome that is a function of various failure modes, and the surfacing of these failure modes is a function of a multidimensional input (various factors defining the environment of use), it is not surprising that a one-dimensional outcome—system reliability—expressed as a function of a one-dimensional input (time on test) is sometimes an incomplete summary.

The next two sections look at common DoD models for reliability growth and at DoD applications of growth models. The discussion in these two sections addresses analytical objectives, underlying assumptions, and practical implementation and interpretation concerns.

COMMON DoD MODELS

Two reliability growth models are used in a majority of current DoD applications: one is a system-level nonhomogeneous Poisson process model with a particular specification of a time-varying intensity function $\lambda(T)$; the other is a competing risk model in which the TAAF program finds and eliminates or reduces failure modes, the remaining risk is reduced, and reliability grows.

The first model is the nonhomogeneous Poisson process formulation[6] with a particular specification of a time-varying intensity function $\lambda(T)$.

---

[5]Software failure modes conceptually can be viewed as deterministic, in the sense that there is no randomness associated with how an element of code acts when called on to support a specific operation. The code either will work as intended or it will "fail," and it repeatedly will demonstrate the identical response each time it is called on to support that same function. The pace at which the code is called on to respond to specific types of operations can, of course, be viewed as random—thereby inducing randomness in software failure processes.

[6]The characterizing feature of this class of models is that the numbers of failures in nonoverlapping time intervals are independent Poisson distributed random variables. A key defining metric for the models is the intensity function $\lambda(T)$ (also referred to as the rate of occurrence of failure). A physically understandable and easily estimable quantity is the cumulative intensity function, $\Lambda(T)$, defined to be $\lambda(T)$ integrated over the time interval $[0, T]$. $\Lambda(T)$ equals the expected cumulative number of failures at time T, that is $\Lambda(T) = E[N(T)]$.

This widely used model, referred to as the power law model,[7] is routinely invoked as the industry standard reliability growth model in DoD acquisition settings. In this model, the failure rate is the following function of T, the cumulative time on test:

$$\lambda(T) = \mu\beta T^{(\beta-1)}, \mu > 0, \beta > 0.$$

This model can be interpreted as a stochastic representation of Duane's postulate (see Crow, 1974) in which the $\log(\lambda(T))$ is a linear function of $\log(T)$. The parameter μ is a scale parameter, while the parameter β determines the degree of reliability growth ($\beta < 1$) or decay ($\beta > 1$). When $\beta = 1$, the model reduces to the homogeneous Poisson process model.

The power law model and various associated statistical methodologies were popularized by the U.S. Army Materiel Systems Analysis Activity (AMSAA), building on Crow (1974) and many other reports (see U.S. Department of Defense, 2011b). Indeed, the power law model is commonly referred to as the AMSAA model, the Crow model, or the AMSAA-Crow model.[8] This continuous reliability growth formulation has been extended to accommodate one-shot reliability data by treating "failure probability" in a manner that parallels that of "failure intensity" in the context of a nonhomogeneous Poisson process, and the "learning curve property" structure is imposed to establish an assumed pattern of reliability growth.

The power law model is a simple analytical representation that facilitates various analytic and inferential actions (e.g., point estimation, confidence bound constructions, and goodness-of-fit procedures). It has also spawned a number of practical follow-on methods for addressing important test program and acquisition oversight issues (see below).

Although it has such practical uses, there are theoretical problems with the power law model. One problem is that the growth in reliability is taken to be continuous over time— increasing while testing is in progress (when no changes to system reliability designs are made) and adhering to the assumed mathematical form when transitioning from one test phase

---

[7]The power law model can be used to represent the reliability of bad as old systems, as in Ascher (1968).

[8]Less common now is the nomenclature Weibull process model, originally motivated by the observation that the intensity function $\lambda(T)$ for the power law model coincides with the form of the failure rate function for the time-to-failure Weibull distribution. The Weibull distribution, however, is not pertinent to this reliability growth setting. For instance, at the end of reliability growth testing under the power law construct, the governing system time-to-failure distribution for future system operations, at and beyond the cumulative test time T, is exponential with a constant mean given by the reciprocal of $\lambda(T)$.

to another (when reliability design enhancements that are implemented provide substantive step-function upgrades to system reliability).[9]

The second reliability growth model more recently used in DoD is based on the assumption that there are a large number of failure modes and that each failure mode operates independently and causes system failure at its own rate. In this model, when a failure mode is observed in testing and subsequently removed by system design enhancements, then the failure rate is reduced at a specific discreet point in time (not continuously, as in the Crow model). If the fix does not introduce new failure modes, reliability grows as a step function.[10] To unify the probabilistic behavior of failure modes prior to corrective actions, additional assumptions can be imposed. For example, for one-shot systems, it is convenient to portray failure mode performance using Bernoulli distributions with associated success probabilities drawn from a common Beta distribution (Hall et al., 2010). Likewise, failure modes for continuously operating systems can be taken to be governed by exponential distributions with failure rates drawn from a parent Gamma distribution (Ellner and Hall, 2006).[11]

Failure modes that are discovered through testing are categorized as either Type A or Type B, corresponding, respectively, to those for which corrective actions will not or will be undertaken (often because of cost or feasibility prohibitions). For each implemented reliability enhancement, the corresponding failure rate or failure probability is assumed to be reduced by some known fix effectiveness factor, which is based on inputs from subject-matter experts or historical data. Although the number of distinct failure modes is unknown, tractable results have been obtained by considering the limit as this count is allowed to approach infinity.

The power law and failure mode-removal models can be viewed as convenient frameworks that facilitate the application of statistical methods to the analysis of reliability test data and the evaluation of reliability testing programs. But they should not be arbitrarily mandated or capriciously imposed. Due consideration needs to be given to the plausibility of underlying assumptions, the possibilities for nonconforming reliability testing data

[9]Sen and Bhattachrayya (1993) developed a more plausible reliability growth model that is consistent with the "learning curve property" but allows reliability to increase only in discrete steps when system design improvements are instituted.

[10]Only one of these fundamental assumptions, statistical independence, is invoked in two failure discount estimation schemes introduced by Lloyd (1987) and used to assess system reliability for certain classes of DoD missiles. Simulation studies, however, indicate that these estimators are strongly positively biased, especially when true system reliability is increasing only modestly during a testing program (Drake, 1987; Fries and Sen, 1996).

[11]For a specific extension of the methodologies based on the primary power law process, Crow (1983) captures the effect of unobserved failure modes by assuming that a second power law representation governs the first times to failure for all individual failure modes, both observed and unobserved.

(especially given any variances in testing circumstances), and the potential sensitivities of analytical results and conclusions.

DoD APPLICATIONS

Reliability growth models can be used to plan the scope of developmental tests, specifically, how much testing time should be devoted to provide a reasonable opportunity for the system design to mature sufficiently in developmental testing (U.S. Department of Defense, 2011b, Ch. 5). Intuitively, key factors in such a determination should include the reliability goal to be achieved by the end of developmental testing (say, R_G), the anticipated initial system reliability at the beginning of developmental testing (say, R_I), and the rate of growth during developmental testing.

The structure of power law formulations directly embraces the growth parameter component of this conceptualization, but the limits of the failure intensity function, especially approaching $T = 0$, do not coincide with physical reality. Nonetheless, with the benefit of a mathematically convenient constraint imposed on the initial phase of testing, the power law process provides a mechanism for assessing the total number of developmental testing hours needed for growing system reliability from R_I to R_G. Simulation-based extensions of the methodology support quantifications of what level for T is required to demonstrate the attainment of the system reliability target R_G with prescribed level of statistical confidence. Other extensions accommodate testing that is focused on individual subsystems (growth or no growth),[12] incorporating analytical aggregations to quantify reliability, and statistical confidence at the system level.

A drawback to this approach is that programmatic risks are sensitive to the length of the first developmental testing event. These shortcomings are not shared by the test planning methodology that is based on the examination of individual failure modes (see U.S. Department of Defense, 2005). That methodology relies on planning parameters that can be directly influenced by program management, such as the fraction of the initial failure rate, or failure probability, addressable by corrective actions (i.e., the management strategy); the average fix effectiveness factor; and the average delay associated with the implementation of corrective actions.

Figure 4-2 displays a typical planning curve (PM-2) for an examination of individual failure modes, highlighting inputs and illustrating key fea-

[12]Testing and analysis at the subsystem level can be appropriate when system functionality is added in increments over time, when opportunities for full-up system testing are limited, and when end-to-end operational scenarios are tested piecemeal in segments or irregularly. Such aggregations, however, need to be carefully scrutinized, especially for deviations from nominal assumptions and effects on robustness.

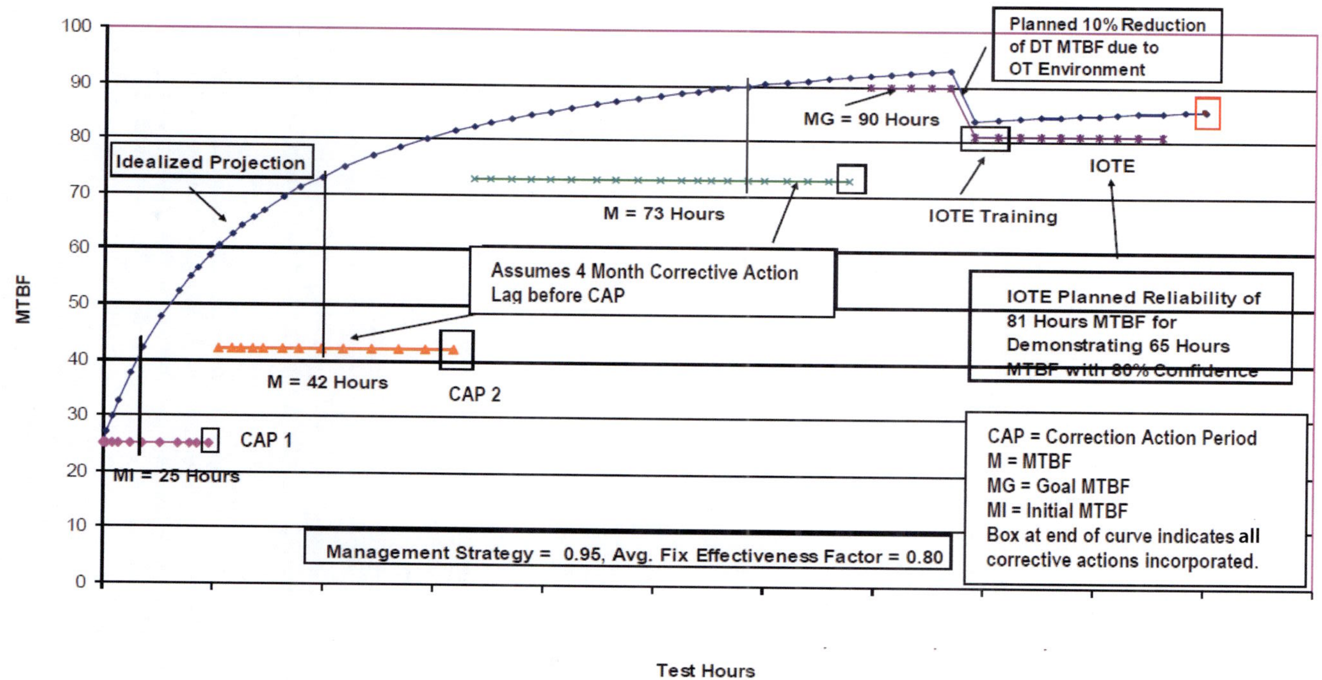

FIGURE 4-2 PM-2 reliability growth planning curve.
NOTES: DT = developmental testing; IOTE = initial test operation evaluation; OT = operational testing; MTBF = mean time between failures. See text for discussion.
SOURCE: U.S. Department of Defense (2011a, Figure 27).

tures.[13] The idealized projection curve is an artificial construct that assumes all observed B-mode failures, those identified for correction, are immediately subjected to fixes. It is transformed to system reliability targets for individual developmental testing events. The number of these events and the respective allocation of testing hours across individual events are variables that planners can adjust. A corrective action period ordinarily follows each testing phase, during which reliability design improvements are implemented.[14] A nominal lag period precedes each such period, in accordance with the notion that the occurrences of B-mode failures near the end of a test phase will not offer adequate time for diagnosis and redesign efforts. The final developmental testing reliability goal (in Figure 4-2, 90 hours mean time between failures) is higher than the assumed operational reliability of the initial operational test and evaluation (81 hours mean time between operational mission failures or a 10 percent reduction). This difference can accommodate potential failure modes that are unique to operational testing (sources of the developmental test/operational test [DT/OT] gap). Likewise, the planned value for the initial operational test and evaluation operational reliability is higher than the operational reliability requirement (65 hours for mean time between operational mission failures), providing some degree of confidence that the requirement will be demonstrated by the initial operational test and evaluation (reducing the consumer's risk).

We note that the PM-2 model is currently mandated for use as a result of a memorandum issued June 26, 2011, "Improving the Reliability of U.S. Army Materiel Systems,"[15] which states

> Program Managers (PMs) of all Acquisition Category I (ACAT I) systems and for ACAT II systems where the sponsor has determined reliability to be an attribute of operational importance shall place reliability growth planning curves in the SEP, TEMP, and Engineering and Manufacturing (EMD) contracts and ensure that U.S. Army systems are resourced to accomplish this requirement. . . . Reliability growth is quantified and reflected through a reliability growth planning curve using the Planning Model based on Projection Methodology (PM2). . . . Where warranted by unique system characteristics, the Army Test and Evaluation Command

---

[13]We note that Figure 4-2 and the preceding discussions treat "reliability" in the general sense, simultaneously encompassing both continuous and discrete data cases (i.e., both those based on mean time between failures and those based on success probability-based metrics). For simplicity, the subsequent exposition in the remainder of this chapter generally will focus on those based on mean time between failures, but parallel structures and similar commentary pertain to systems that have discrete performance.

[14]Not all corrective actions are implemented following a test period; some require longer time periods for development and incorporation.

[15]The document is available at http://www.amsaa.army.mil/Documents/Reliability%20Policy_6_29_2011_13_10_10.pdf [August 2014].

(ATEC), in consultation with the Project Manager (PM), may specify an alternative reliability growth planning method.

A reliability growth planning curve is important for developing an overall strategy for testing and evaluation, for defining individual testing events and determining requisite supporting resources, and for providing a series of reliability targets that can be tracked to judge the progress of reliability growth. The Director of Operational Test and Evaluation (DOT&E) requires that a reliability growth curve appear in the system's Test and Evaluation Master Plan (TEMP), but does not prescribe the specific mechanism by which the plan is to be developed. As program milestones are achieved or in response to unanticipated testing outcomes, the reliability growth curve, as well as the entire TEMP, is expected to be updated.

IMPLICATIONS

The DOT&E requirement for presenting and periodically revising a formal reliability growth planning curve is eminently reasonable. To generate its curve, the responsible program office can follow existing standard methods or use other approaches it deems suitable.[16] What is important in practice is that any proposed reliability growth curve is fully integrated with the overall system development and test and evaluation strategies (e.g., accommodating other performance issues not related to reliability), recognizes potential sensitivities to underlying analytical assumptions, and retains adequate flexibility to respond to emerging testing results.

There are three key elements of a reliability growth curve that warrant emphasis. First, it should provide a mechanism for early checks of the adequacy of system design for reliability. Second, rough adherence to the planning curve should position a developmental program so that the initial operational test and evaluation, as a stand-alone test, will demonstrate the attainment of the operational reliability requirement with high confidence. Third, since the construction of a planning curve rests on numerous assumptions, some of which may turn out to be incompatible with the subsequent testing experience, sensitivity and robustness of the modeling need to be understood and modifications made when warranted.

Once a developmental program begins system-level testing, reliability growth methodologies are available for estimating model parameters, constructing curves that portray how demonstrated reliabilities have evolved

[16]In the extreme, given a general sense of the developmental testing time available for a particular system and the customary nature of development tests ordinarily undertaken for such classes of systems, one could imagine divining a simple eye-ball fit through a potentially suitable smooth curve that traces from R_I to some established mark above R_G.

and compare to planned trajectories, and projecting system reliability estimates beyond what has been achieved to date (U.S. Department of Defense, 2011b, Ch. 6). There is a natural inclination for reliability analysts to routinely invoke these methods, especially when faced with budget constraints and schedule demands that cry for "efficiencies" in testing and evaluation by using all of the available data. Likewise, there is an instinctive desire for program management and oversight agencies to closely monitor a program's progress and to support decisions backed by "high confidence" analyses. In both settings, reliability growth methods *offer* the prospects of accessible data synthesis—directly through simple equations or by the application of dedicated software packages.

It is sensible to view a reliability growth methodology as a potential tool for supporting in-depth assessments of system reliability, but it should not be assumed in advance to be the single definitive mechanism underpinning such analyses. Comprehensive evaluations of system reliability would consider the spectrum of testing circumstances and their relation to operational profiles, exhibited failure modes and rates for individual test articles and collections (rather than merely system failure times), insights from reliability and design engineers, observations from system operators, and a multiple of other potentially important factors germane to the specific system under development. Subsequently, after due diligence, it may be determined that standard reliability growth methods provide a reasonable approach for addressing a specific analytical issue or for conveniently portraying bottom-line conclusions.

There are a number of reasons that reliability results recorded over the course of developmental testing may not match target values or thresholds prescribed in advance by the associated reliability growth planning curve. Not all of these differences should translate to alarms that system reliability is problematic or deficient, nor should one assume that close conformity of developmental testing results to a reliability planning curve by itself ensures the adequacy of system operational reliability (e.g., the developmental tests may be unrepresentative of more stressful operational circumstances). Again, a detailed understanding of the testing conditions and the extent of divergences from operationally realistic environments and use is critical for meaningful interpretation.

Benign system-level tests (e.g., some types of laboratory or chamber testing common to early developmental testing) may yield inflated reliability results. When followed by testing characterized by more realistic loads and stresses, the apparent trend may suggest that system reliability is in decline. Similar deviations, either upwards or downwards, may be evident in the midst of a developmental testing program when the severity of the environments and use profiles is changed from test to test. Even when the intent is for uniform testing under a prescribed profile of use, test-specific irregularities can be anticipated. Crow (2008) presents a method for checking the consistency of use profiles at intermediate pre-determined "convergence points" (expressed in terms of accumulated testing time, vehicle mileage, cycles completed, etc.) and accordingly adjusting planned follow-on testing.

When system functionality is added incrementally (e.g., software modules are added as they are developed), testing of the more advanced system configurations may exhibit a relative degradation in system reliability—primarily if the unreliability of the new aspect of the system dominates the enhancements that had been incorporated in response to observations from the immediately preceding test event. Similar effects are possible when a system operationally interfaces with external systems (i.e., an element that is not part of the subject system under development), one of those peripheral systems is modified (perhaps as part of its own developmental cycles), and interface "failures" formally chargeable against the subject system occur in follow-on testing of the subject system.

There are situations in which the demonstrated system reliability falls far short of what was planned, and, after careful review, the likely candidate for the disappointing performance is some combination of an initially deficient reliability design and an inadequate reliability enhancement program. For example, the number of system reliability failures recorded in a system's first developmental testing event may be well beyond what was envisioned for that initial test or even may far exceed what was planned for the entire developmental testing program (i.e., the complete set of planned tests). Unfortunately, this has been a too common outcome in the recent history of DoD reliability testing. Another disturbing situation is that after a few test events reliability estimates stagnate well below targeted values, while the counts of new failure modes continue to increase.

A visually detectable major departure from the planning curve by itself could provide a triggering mechanism for instituting an in-depth program review. Supporting statistical evidence can be provided by constructed confidence bounds and associated hypothesis tests, or by a formal assessment of the estimated remaining "growth potential" for the system's reliability (deducing the theoretical upper limit for the system reliability that could be obtained within the defined testing program). The growth potential calculation could indicate that there is little chance for the program to succeed unless major system redesigns and institutional reliability management procedures are implemented (i.e., essentially constituting a new reliability growth program). Alternatively, it could show that there is no strong evidence that would compel a program intervention.

Determining the system's reliability growth potential is a type of forecasting, functionally extrapolating forward to the ultimate attainable limit. Another documented projection methodology is essentially a one-step-ahead estimate of the system reliability—taking the level of reliability demonstrated by completed reliability growth testing and giving additional credit for the last set of implemented reliability improvements. These approaches rely on values of failure-mode-specific fix effectiveness factors provided by subject-matter experts. If the effects of the unobserved failure modes are ignored, then positively biased estimates are produced. There are at present no associated statistical confidence bound procedures for this situation, but the setting would seem to be amenable to the application of bootstrap and Bayesian techniques.

Projection-based estimates of system reliability offer a potential recourse when the conducted growth testing indicates that the achieved reliability falls short of a critical programmatic mark. If the shortfall is significant, then the inherent subjectivity and uncertainty of provided fix effectiveness factors naturally limits the credibility of a projection-based "demonstration" of compliance. Supplementary independent reliability engineering analyses, dedicated confirmatory technical testing, or follow-on system-level testing may be warranted.

This description of the current state of reliability growth modeling highlights some issues concerning the validity of these models. Two key concerns are that time on test is often not a good predictor linking time with system reliability, and that reliability growth models generally fail to represent the test circumstances. These two criticisms raise concerns about the applicability of these models.

As noted above, DoD currently uses reliability growth models for three important purposes. First, prior to development, they are used to budget the number of hours of testing needed to grow the reliability from what it is conjectured to be as the result of initial design work to the level of the operational requirement. Because so little is known about the system at that early point in development, one cannot specify a clearly superior method: thus, the use of reliability growth models in this role is reasonable—provided that the parameters central to the specific invoked reliability growth model are plausibly achievable (e.g., based on experiences with similar classes of systems). In the case of nonhomogeneous Poisson process models, a good approach would be to select the growth parameter, β, by examining the growth observed for systems with similar structure, including some components that are essentially identical that have already been through developmental and operational testing. The panel could not determine whether this approach is currently commonplace.

Second, during development, reliability growth models are used to combine reliability assessments over test events to track the current level of reliability attained. Again, for this application, reliability growth methodologies are appropriate—subject to the validation of inherent model assumptions. However, it seems as though in many cases reliability growth models serve merely as curve fitting mechanisms. In this circumstance, we doubt that reliability growth models would be found to be clearly superior to straightforward regression or time-series approaches.

Third, reliability growth models offer forecasting capabilities—to predict either the time at which the required reliability level ultimately will be attained or the reliability to be realized at a specific time. Here, the questions concerning the validity of reliability growth models are of the greatest concern because extrapolation is a more severe test than interpolation. Consequently, the panel does not support the use of these models for such predictions, absent a comprehensive validation. If such a validation is carried out, then the panel thinks it is likely that it will regularly demonstrate the inability of such models to predict system reliability past the very near future.

5

System Design for Reliability

Design for reliability is a collection of techniques that are used to modify the initial design of a system to improve its reliability. It appears to the panel that U.S. Department of Defense (DoD) contractors do not fully exploit these techniques. There are probably a variety of reasons for this omission, including the additional cost and time of development needed. However, such methods can dramatically increase system reliability, and DoD system reliability would benefit considerably from the use of such methods. This chapter describes techniques to improve system design to enhance system reliability.

From 1980 until the mid-1990s, the goal of DoD reliability policies was to achieve high initial reliability by focusing on reliability fundamentals during design and manufacturing. Subsequently, DoD allowed contractors to rely primarily on "testing reliability in" toward the end of development. This change was noted in the 2011 *Annual Report to Congress* of the Director of Operational Test and Evaluation (U.S. Department of Defense, 2011b, p. v):

> [I]ndustry continues to follow the 785B methodology, which unfortunately takes a more reactive than proactive approach to achieving reliability goals. In this standard, approximately 30 percent of the system reliability comes from the design while the remaining 70 percent is to be achieved through growth implemented during the test phases.

This pattern points to the need for better design practices and better system engineering (see also Trapnell, 1984; Ellner and Trapnell, 1990).

Many developers of defense systems depend on reliability growth methods applied after the initial design stage to achieve their required levels of reliability. Reliability growth methods, primarily utilizing test-analyze-fix-test, are an important part of nearly any reliability program, but "testing reliability in" is both inefficient and ineffective in comparison with a development approach that uses design-for-reliability methods. Relying on testing-in reliability is inefficient and ineffective because when failure modes are discovered late in system development, corrective actions can lead to delays in fielding and cost over-runs in order to modify the system architecture and make any related changes. In addition, fixes incorporated late in development often cause problems in interfaces, because of a failure to identify all the effects of a design change, with the result that the fielded system requires greater amounts of maintenance and repair.

Traditional military reliability prediction methods, including those detailed in *Military Handbook: Reliability Prediction of Electronic Equipment* (MIL-HDBK-217) (U.S. Department of Defense, 1991), rely on the collection of failure data and generally assume that the components of the system have failure rates (most often assumed to be constant over time) that can be modified by independent "modifiers" to account for various quality, operating, and environmental conditions. MIL-HDBK-217, for example, offers two methods for predicting reliability, the "stress" method and the "parts count" method. In both of these methods, a generic average failure rate (assuming average operating conditions) is assumed. The shortcoming of this approach is that it uses only the field data, without understanding the root cause of failure (for details, see Pecht and Kang, 1988; Wong, 1990; Pecht et al., 1992). This approach is inaccurate for predicting actual field failures and provides highly misleading predictions, which can result in poor designs and logistics decisions.

An emerging approach uses physics-of-failure and design-for-reliability methods (see, e.g., Pecht and Dasgupta, 1995). Physics of failure uses knowledge of a system's life-cycle loading and failure mechanisms to perform reliability modeling, design, and assessment. The approach is based on the identification of potential failure modes, failure mechanisms, and failure sites for the system as a function of its life-cycle loading conditions. The stress at each failure site is obtained as a function of both the loading conditions and the system geometry and material properties. Damage models are used to determine fault generation and propagation.

Many reliability engineering methods have been developed and are collectively referred to as design for reliability (a good description can be found in Pecht, 2009). Design for reliability includes a set of techniques that support the product design and the design of the manufacturing process that greatly increase the likelihood that the reliability requirements are met throughout the life of the product with low overall life-cycle costs. The techniques that comprise design for reliability include (1) failure modes and effects analysis, (2) robust parameter design, (3) block diagrams and fault tree analyses, (4) physics-of-failure methods, (5) simulation methods, and (6) root-cause analysis. Over the past 20 years, manufacturers of many commercial products have learned that to expedite system development and to contain costs (both development costs and life-cycle or warranty costs) while still meeting or exceeding reliability requirements, it is essential to use modern design-for-reliability tools as part of a program to achieve reliability requirements.

In particular, physics of failure is a key approach used by manufacturers of commercial products for reliability enhancement. While traditional reliability assessment techniques heavily penalize systems making use of new materials, structures, and technologies because of a lack of sufficient field failure data, the physics-of-failure approach is based on generic failure models that are as effective for new materials and structures as they are for existing designs. The approach encourages innovative designs through a more realistic reliability assessment.

The use of design-for-reliability techniques can help to identify the components that need modification early in the design stage when it is much more cost-effective to institute such changes. In particular, physics-of-failure methods enable developers to better determine what components need testing, often where there remains uncertainty about the level of reliability in critical components.

A specific approach to design for reliability was described during the panel's workshop by Guangbin Yang of Ford Motor Company. Yang said that at Ford they start with the design for a new system, which is expressed using a system boundary diagram along with an interface analysis. Then design mistakes are discovered using computer-aided engineering, design reviews, failure-mode-and-effects analysis, and fault-tree analysis. Lack of robustness of designs is examined through use of a P-diagram, which examines how noise factors, in conjunction with control factors and the anticipated input signals, generate an output response, which can include various errors.

We emphasize throughout this report the need for assessment of full-system reliability. In addition, at this point in the development process, there would also be substantial benefits of an assessment of the reliability of high-cost and safety critical subsystems for both the evaluation of the current system reliability and the reliability of future systems with similar subsystems. Such a step is almost a prerequisite of assessment of full-system reliability.

TECHNIQUES FOR DESIGN

Producing a reliable system requires planning for reliability from the earliest stages of system design. Assessment of reliability as a result of design choices is often accomplished through the use of probabilistic design for reliability, which compares a component's strength against the stresses it will face in various environments. These practices can substantially increase reliability through better system design (e.g., built-in redundancy) and through the selection of better parts and materials. In addition, there are practices that can improve reliability with respect to manufacturing, assembly, shipping and handling, operation, maintenance and repair. These practices, collectively referred to as design for reliability, improve reliability through design in several ways:

- They ensure that the supply-chain participants have the capability to produce the parts (materials) and services necessary to meet the final reliability objectives and that those participants are following through.
- They identify the potential failure modes, failure sites, and failure mechanisms.
- They design to the quality level that can be controlled in manufacturing and assembly, considering the potential failure modes, failure sites, and failure mechanisms, obtained from the physics-of-failure analysis, and the life-cycle profile.
- They verify the reliability of the system under the expected life-cycle conditions.
- They demonstrate that all manufacturing and assembly processes are capable of producing the system within the statistical process window required by the design. Because variability in material properties and manufacturing processes will affect a system's reliability, characteristics of the process must be identified, measured, and monitored.
- They manage the life-cycle usage of the system using closed loop, root-cause monitoring procedures.

Reviewing in-house procedures (e.g., design, manufacturing process, storage and handling, quality control, maintenance) against corresponding standards can help identify factors that could cause failures. For example, misapplication of a component could arise from its use outside the operating conditions specified by the vendor (e.g., current, voltage, or temperature). Equipment misapplication can result from improper changes in the operating requirements of the machine.

After these preliminaries, once design work is initiated, the goal is to determine a design for the system that will enable it to have high initial reliability prior to any formal testing. Several techniques for design for reliability are discussed in the rest of this section: defining and characterizing life-cycle loads to improve design parameters; proper selection of parts and materials; and analysis of failure modes, mechanisms, and effects.

Defining and Characterizing Life-Cycle Loads

The life-cycle conditions of any system influence decisions concerning: (1) system design and development, (2) materials and parts selection, (3) qualification, (4) system safety, and (5) maintenance. The phases in a system's life cycle include manufacturing and assembly, testing, rework, storage, transportation and handling, operation, and repair and maintenance (for an example of the impact on reliability of electronic components as a result of shock and random vibration life-cycle loads, see Mathew et al., 2007). During each phase of its life cycle, a system will experience various environmental and usage stresses. The life-cycle stresses can include, but are not limited to: thermal, mechanical (e.g., pressure levels and gradients, vibrations, shock loads, acoustic levels), chemical, and electrical loading conditions. The degree of and rate of system degradation, and thus reliability, depend upon the nature, magnitude, and duration of exposure to such stresses.

Defining and characterizing the life-cycle stresses can be difficult because systems can experience completely different application conditions, including location, the system utilization profile, and the duration of utilization and maintenance conditions. In other words, there is no precise description of the operating environment for any system.[1] Consider the example of a computer, which is typically designed for a home or office environment. However, the operational profile of each computer may be completely different depending on user behavior. Some users may shut down the computer every time they log off; others may shut down only once at the end of the day; still others may keep their computers on all the time. Furthermore, one user may keep the computer by a sunny window, while another person may keep the computer nearby an air conditioner, so the temperature profile experienced by each system, and hence its degradation due to thermal loads, would be different.

There are three methods used to estimate system life-cycle loads relevant to defense systems: similarity analysis, field trial and service records, and in-situ monitoring:

---

[1]This is one of the limitations of prediction that is diminishing over time, given that many systems are being outfitted with sensors and communications technology that provide comprehensive information about the factors that will affect reliability.

- Similarity analysis estimates environmental stresses when sufficient field histories for similar systems are available. Before using data on similar systems for proposed designs, the characteristic differences in design and application for the comparison systems need to be reviewed. For example, electronics inside a washing machine in a commercial laundry are expected to experience a wider distribution of loads and use conditions (because of a large number of users) and higher usage rates than a home washing machine.
- Field trial records provide estimates of the environmental profiles experienced by the system. The data are a function of the lengths and conditions of the trials and can be extrapolated to estimate actual user conditions. Service records provide information on the maintenance, replacement, or servicing performed.
- In-situ monitoring (for a good example, see Das, 2012) can track usage conditions experienced by the system over a system's life cycle. These data are often collected using sensors. Load distributions can be developed from data obtained by monitoring systems that are used by different users. The data need to be collected over a sufficiently long period to provide an estimate of the loads and their variation over time. In-situ monitoring provides the most accurate account of load histories and is most valuable in design for reliability.

Proper Selection of Parts and Materials

Almost all systems include parts (materials) produced by supply chains of companies. It is necessary to select the parts (materials) that have sufficient quality and are capable of delivering the expected performance and reliability in the application. Because of changes in technology trends, the evolution of complex supply-chain interactions and new market challenges, shifts in consumer demand, and continuing standards reorganization, a cost-effective and efficient parts selection and management process is needed to perform this assessment, which is usually carried out by a multidisciplinary team. (For a description of this process for an electronic system, see Sandborn et al., 2008.) A manufacturer's ability to produce parts with consistent quality is evaluated; the distributor assessment evaluates the distributor's ability to provide parts without affecting the initial quality and reliability; and the parts selection and management team defines the minimum acceptability criteria based on a system's requirements.

In the next step, the candidate part is subjected to application-dependent assessments. The manufacturer's quality policies are assessed with respect to five assessment categories: process control; handling, storage, and shipping controls; corrective and preventive actions; product traceability; and change notification. If the part is not found to be acceptable after this assessment, then the assessment team must decide whether an acceptable alternative is available. If no alternative is available, then the team may choose to pursue techniques that mitigate the possible risks associated with using an unacceptable part.

Performance assessment seeks to evaluate a part's ability to meet the performance requirements (e.g., functional, mechanical, and electrical) of the system. In order to increase performance, manufacturers may adopt features for products that make them less reliable.

In general, there are no distinct boundaries for such stressors as mechanical load, current, or temperature above which immediate failure will occur and below which a part will operate indefinitely. However, there are often a minimum and a maximum limit beyond which the part will not function properly or at which the increased complexity required to address the stress with high probability will not offer an advantage in cost-effectiveness. The ratings of the part manufacturer or the user's procurement ratings are generally used to determine these limiting values. Equipment manufacturers who use such parts need to adapt their design so that the part does not experience conditions beyond its ratings. It is the responsibility of the parts team to establish that the electrical, mechanical, or functional performance of the part is suitable for the life-cycle conditions of the particular system.

Failure Modes, Mechanisms, and Effects Analysis

A failure mode is the manner in which a failure (at the component, subsystem, or system level) is observed to occur, or alternatively, as the specific way in which a failure is manifested, such as the breaking of a truck axle. Failures do link hierarchically in terms of the system architecture, and so a failure mode may, in turn, cause failures in a higher level subsystem or may be the result of a failure of a lower level component, or both. A failure cause is defined as the circumstances during design, manufacture, storage, transportation, or use that lead to a failure. For each failure mode, there may be many potential causes that can be identified.

Failure mechanisms are the processes by which specific combinations of physical, electrical, chemical, and mechanical stresses induce failure. Failure mechanisms are categorized as either overstress or wear-out mechanisms; an overstress failure involves a failure that arises as a result of a single load (stress) condition. Wear-out failure involves a failure that arises as a result of cumulative load (stress) conditions. Knowledge of the likely failure mechanisms is essential for developing designs for reliable systems.

Failure modes, mechanisms, and effects analysis is a systematic approach to identify the failure mechanisms and models for all potential failure modes, and to set priorities among them. It supports physics-of-failure-based design for reliability. High-priority failure mechanisms determine the operational stresses and the environmental and operational parameters that need to be accounted or controlled for in the design.

Failure modes, mechanisms, and effects analysis is used as input in the determination of the relationships between system requirements and the physical characteristics of the product (and their variation in the production process), the interactions of system materials with loads, and their influences on the system's susceptibility to failure with respect to the use conditions. This process merges the design-for-reliability approach with material knowledge. It uses application conditions and the duration of the application with understanding of the likely stresses and potential failure mechanisms. The potential failure mechanisms are considered individually, and they are assessed with models that enable the design of the system for the intended application.

Failure models use appropriate stress and damage analysis methods to evaluate susceptibility of failure. Failure susceptibility is evaluated by assessing the time to failure or likelihood of a failure for a given geometry, material construction, or environmental and operational condition. Failure models of overstress mechanisms use stress analysis to estimate the likelihood of a failure as a result of a single exposure to a defined stress condition. The simplest formulation for an overstress model is the comparison of an induced stress with the strength of the material that must sustain that stress.

Wear-out mechanisms are analyzed using both stress and damage analysis to calculate the time required to induce failure as a result of a defined stress life-cycle profile. In the case of wear-out failures, damage is accumulated over a period until the item is no longer able to withstand the applied load. Therefore, an appropriate method for combining multiple conditions has to be determined for assessing the time to failure. Sometimes, the damage due to the individual loading conditions may be analyzed separately, and the failure assessment results may be combined in a cumulative manner.

Life-cycle profiles include environmental conditions such as temperature, humidity, pressure, vibration or shock, chemical environments, radiation, contaminants, and loads due to operating conditions, such as current, voltage, and power. The life-cycle environment of a system consists of assembly, storage, handling, and usage conditions of the system. Information on life-cycle conditions can be used for eliminating failure modes that may not occur under the given application conditions.

In the absence of field data, information on system use conditions can be obtained from environmental handbooks or from data collected on similar environments. Ideally, such data should be obtained and processed during actual application. Recorded data from the life-cycle stages for the same or similar products can serve as input for a failure modes, mechanisms, and effects analysis.

Ideally all failure mechanisms and their interactions are considered for system design and analysis. In the life cycle of a system, several failure mechanisms may be activated by different environmental and operational parameters acting at various stress levels, but only a few operational and environmental parameters and failure mechanisms are in general responsible for the majority of the failures (see Mathew et al., 2012). High-priority mechanisms are those that may cause the product to fail relatively early in a product's intended life. These mechanisms occur during the normal operational and environmental conditions of the product's application.

Failure susceptibility is evaluated using the previously identified failure models when they are available. For overstress mechanisms, failure susceptibility is evaluated by conducting a stress analysis under the given environmental and operating conditions. For wear-out mechanisms, failure susceptibility is evaluated by determining the time to failure under the given environmental and operating conditions. If no failure models are available, then the evaluation is based on past experience, manufacturer data, or handbooks.

After evaluation of failure susceptibility, occurrence ratings under environmental and operating conditions applicable to the system are assigned to the failure mechanisms. For the overstress failure mechanisms that precipitate failure, the highest occurrence rating, "frequent," is assigned. If no overstress failures are precipitated, then the lowest occurrence rating, "extremely unlikely," is assigned. For the wear-out failure mechanisms, the ratings are assigned on the basis of benchmarking the individual time to failure for a given wear-out mechanism with overall time to failure, expected product life, past experience, and engineering judgment.

The purpose of failure modes, mechanisms, and effects analysis is to identify potential failure mechanisms and models for all potential failures modes and to prioritize them. To ascertain the criticality of the failure mechanisms, a common approach is to calculate a risk priority number for each mechanism. The higher the risk priority number, the higher a failure mechanism is ranked. That number is the product of the probability of detection, occurrence, and severity of each mechanism. Detection describes the probability of detecting the failure modes associated with the failure mechanism. Severity describes the seriousness of the effect of the failure caused by a mechanism. Additional insights into the criticality of a failure mechanism can be obtained by examining past repair and maintenance actions, the reliability capabilities of suppliers, and results observed in the initial development tests.

TECHNIQUES TO ASSESS RELIABILITY POTENTIAL

Assessment of the reliability potential of a system design is the determination of the reliability of a system consistent with good practice and conditional on a use profile. The reliability potential is estimated through use of various forms of simulation and component-level testing, which include integrity tests, virtual qualification, and reliability testing.

Integrity Tests

Integrity is a measure of the appropriateness of the tests conducted by the manufacturer and of the part's ability to survive those tests. Integrity test data (often available from the part manufacturer) are examined in light of the life-cycle conditions and applicable failure mechanisms and models. If the magnitude and duration of the life-cycle conditions are less severe than those of the integrity tests, and if the test sample size and results are acceptable, then the part reliability is acceptable. If the integrity test data are insufficient to validate part reliability in the application, then virtual qualification should be considered.

Virtual Qualification

Virtual qualification can be used to accelerate the qualification process of a part for its life-cycle environment. Virtual qualification uses computer-aided simulation to identify and rank the dominant failure mechanisms associated with a part under life-cycle loads, determine the acceleration factor for a given set of accelerated test parameters, and determine the expected time to failure for the identified failure mechanisms (for an example, see George et al., 2009).

Each failure model is made up of a stress analysis model and a damage assessment model. The output is a ranking of different failure mechanisms, based on the time to failure. A stress model captures the product architecture, while a damage model depends on a material's response to the applied stress. Virtual qualification can be used to optimize the product design in such a way that the minimum time to failure of any part of the product is greater than its desired life. Although the data obtained from virtual qualification cannot fully replace the data obtained from physical tests, they can increase the efficiency of physical tests by indicating the potential failure modes and mechanisms that can be expected.

Ideally, a virtual qualification process will identify quality suppliers and quality parts through use of physics-of-failure modeling and a risk assessment and mitigation program. The process allows qualification to be incorporated into the design phase of product development, because it allows design, manufacturing, and testing to be conducted promptly and cost-effectively.

The effects of manufacturing variability can be assessed by simulation as part of the virtual qualification process. But it is important to remember that the accuracy of the results using virtual qualification depends on the accuracy of the inputs to the process, that is, the system geometry and material properties, the life-cycle loads, the failure models used, the analysis domain, and the degree of discreteness used in the models (both spatial and temporal). Hence, to obtain a reliable prediction, the variability in the inputs needs to be specified using distribution functions, and the validity of the failure models needs to be tested by conducting accelerated tests (see Chapter 6 for discussion).

Reliability Testing

Reliability testing can be used to determine the limits of a system, to examine systems for design flaws, and to demonstrate system reliability. The tests may be conducted according to industry standards or to required customer specifications. Reliability testing procedures may be general, or the tests may be specifically designed for a given system.

The information required for designing system-specific reliability tests includes the anticipated life-cycle conditions, the reliability goals for the system, and the failure modes and mechanisms identified during reliability analysis. The different types of reliability tests that can be conducted include tests for design marginality, determination of destruct limits, design verification testing before mass production, on-going reliability testing, and accelerated testing (for examples, see Keimasi et al., 2006; Mathew et al., 2007; Osterman 2011; Alam et al., 2012; and Menon et al., 2013).

Many testing environments may need to be considered, including high temperature, low temperature, temperature cycle and thermal shock, humidity, mechanical shock, variable frequency vibration, atmospheric contaminants, electromagnetic radiation, nuclear/cosmic radiation, sand and dust, and low pressure:

- High temperature: High-temperature tests assess failure mechanisms that are thermally activated. In electromechanical and mechanical systems, high temperatures may soften insulation, jam moving parts because of thermal expansion, blister finishes, oxidize materials, reduce viscosity of fluids, evaporate lubricants, and cause structural overloads due to physical expansions. In electrical systems, high temperatures can cause variations in resistance, inductance, capacitance, power factor, and dielectric constant.

- Low temperature: In mechanical and electromechanical systems, low temperatures can cause plastics and rubber to lose flexibility and become brittle, cause ice to form, increase viscosity of lubricants and gels, and cause structural damage due to physical contraction. In electrical systems, low-temperature tests are performed primarily to accelerate threshold shifts and parametric changes due to variation in electrical material parameters.
- Temperature cycle and thermal shock: Temperature cycle and thermal shock testing are most often used to assess the effects of thermal expansion mismatch among the different elements within a system, which can result in materials' overstressing and cracking, crazing, and delamination.
- Humidity: Excessive loss of humidity can cause leakage paths between electrical conductors, oxidation, corrosion, and swelling in materials such as gaskets and granulation.
- Mechanical shock: Some systems must be able to withstand a sudden change in mechanical stress typically due to abrupt changes in motion from handling, transportation, or actual use. Mechanical shock can lead to overstressing of mechanical structures causing weakening, collapse, or mechanical malfunction.
- Variable frequency vibration: Some systems must be able to withstand deterioration due to vibration. Vibration may lead to the deterioration of mechanical strength from fatigue or overstress; may cause electrical signals to be erroneously modulated; and may cause materials and structure to crack, be displaced, or be shaken loose from mounts.
- Atmospheric contaminants: The atmosphere contains such contaminants as airborne acids and salts that can lower electrical and insulation resistance, oxidize materials, and accelerate corrosion. Mixed flowing gas tests are often used to assess the reliability of parts that will be subjected to these environments.
- Electromagnetic radiation: Electromagnetic radiation can cause spurious and erroneous signals from electronic components and circuitry. In some cases, it may cause complete disruption of normal electrical equipment such as communication and measuring systems.
- Nuclear/cosmic radiation: Nuclear/cosmic radiation can cause heating and thermal aging; alter the chemical, physical, and electrical properties of materials; produce gasses and secondary radiation; oxidize and discolor surfaces; and damage electronic components and circuits.
- Sand and dust: Sand and dust can scratch and abrade finished sur-

faces; increase friction between surfaces, contaminate lubricants, clog orifices, and wear materials.

- Low pressure: Low pressure can cause overstress of structures such as containers and tanks that can explode or fracture; cause seals to leak; cause air bubbles in materials, which may explode; lead to internal heating due to lack of cooling medium; cause arcing breakdowns in insulations; lead to the formation of ozone; and make outgassing more likely.

Reliability test data analysis can be used to provide a basis for design changes prior to mass production, to help select appropriate failure models and estimate model parameters, and for modification of reliability predictions for a product. Test data can also be used to create guidelines for manufacturing tests including screens, and to create test requirements for materials, parts, and sub-assemblies obtained from suppliers.

We stress that the still-used handbook MIL-HDBK-217 (U.S. Department of Defense, 1991) does not provide adequate design guidance and information regarding microelectronic failure mechanisms. In many cases, MIL-HDBK-217 methods would not be able to distinguish between separate failure mechanisms. It is in clear contrast with physics-of-failure estimation: "an approach to design, reliability assessment, testing, screening and evaluating stress margins by employing knowledge of root-cause failure processes to prevent product failures through robust design and manufacturing practices" (Cushing et al., 1993, p. 542). A detailed critique of MIL-HDBK-217 is provided in Appendix D.

ANALYSIS OF FAILURES AND THEIR ROOT CAUSES

Failure tracking activities are used to collect test- and field-failed components and related failure information. Failures have to be analyzed to identify the root causes of manufacturing defects and to test or field failures. The information collected needs to include the failure point (quality testing, reliability testing, or field), the failure site, and the failure mode and mechanism. For each product category, a Pareto chart of failure causes can be created and continually updated.

The outputs for this key practice are a failure summary report arranged in groups of similar functional failures, actual times to failure of components based on time of specific part returns, and a documented summary of corrective actions implemented and their effectiveness. All the lessons learned from failure analysis reports can be included in a corrective actions database for future reference. Such a database can help save considerable funds in fault isolation and rework associated with future problems.

A classification system of failures, failure symptoms, and apparent causes can be a significant aid in the documentation of failures and their root causes and can help identify suitable preventive methods. By having such a classification system, it may be easier for engineers to identify and share information on vulnerable areas in the design, manufacture, assembly, storage, transportation, and operation of the system. Broad failure classifications include system damage or failure, loss in operating performance, loss in economic performance, and reduction in safety. Failures categorized as system damage can be further categorized according to the failure mode and mechanism. Different categories of failures may require different root-cause analysis approaches and tools.

The goal of failure analysis is to identify the root causes of failures. The root cause is the most basic causal factor or factors that, if corrected or removed, will prevent the recurrence of the failure. Failure analysis techniques include nondestructive and destructive techniques. Nondestructive techniques include visual observation and observations under optical microscope, x-ray, and acoustic microscopy. Destructive techniques include cross-sectioning of samples and de-capsulation. Failure analysis is used to identify the locations at which failures occur and the fundamental mechanisms by which they occurred. Failure analysis will be successful if it is approached systematically, starting with nondestructive examinations of the failed test samples and then moving on to more advanced destructive examinations; see Azarian et al. (2006) for an example.

Product reliability can be ensured by using a closed-loop process that provides feedback to design and manufacturing in each stage of the product life cycle, including after the product is shipped and fielded. Data obtained from maintenance, inspection, testing, and usage monitoring can be used to perform timely maintenance for sustaining the product and for preventing failures.

An important tool in failure analysis is known as FRACAS or failure reporting, analysis and corrective action system. According to the Reliability Analysis Center:

> A failure reporting, analysis and corrective action system (FRACAS) is defined, and should be implemented, as a closed-loop process for identifying and tracking root failure causes, and subsequently determining, implementing and verifying an effective corrective action to eliminate their reoccurrence. The FRACAS accumulates failure, analysis and corrective action information to assess progress in eliminating hardware, software and process-related failure modes and mechanisms. It should contain information and data to the level of detail necessary to identify design or process deficiencies that should be eliminated.

It is important for FRACAS to be applied throughout developmental and operational testing and post-deployment.

TWO APPROACHES TO RELIABILITY PREDICTION

Reliability predictions are an important part of product design. They are used for a number of different purposes: (1) contractual agreements, (2) feasibility evaluations, (3) comparisons of alternative designs, (4) identification of potential reliability problems, (5) maintenance and logistics support planning, and (6) cost analyses. As a consequence, erroneous reliability predictions can result in serious problems during development and after a system is fielded. An overly optimistic prediction, estimating too few failures, can result in selection of the wrong design, budgeting for too few spare parts, expensive rework, and poor field performance. An overly pessimistic prediction can result in unnecessary additional design and test expenses to resolve the perceived low reliability. This section discusses two explicit models and similarity analyses for developing reliability predictions.

Two Explicit Models

Fault trees and reliability block diagrams are two methods for developing assessments of system reliabilities from those of component reliabilities: see Box 5-1.[2] Although they can be time-consuming and complex (depending on the level of detail applied), they can accommodate model dependencies. Nonconstant failure rates can be handled by assessing the probability of failure at different times using the probability of failure for each component at each time, rather than using the component's mean time between failure. Thus, components can be modeled to have decreasing, constant, or increasing failure rates. These methods can also accommodate time-phased missions. Unfortunately, there may be so many ways to fail a system that an explicit model (one which identifies all the failure possibilities) can be intractable. Solving these models using the complete enumeration method is discussed in many standard reliability text books (see, e.g., Meeker and Escobar (1998); also see *Guide for Selecting and Using Reliability Predictions* of the IEEE Standards Association [IEEE 1413.1]).

[2]For additional design-for-reliability tools that have proven useful in DoD acquisition, see Section 2.1.4 of the TechAmerica Reliability Program Handbook, TA-HB-0009, available: http://www.techstreet.com/products/1855520 [August 2014].

BOX 5-1
Two Common Techniques for Design for Reliability

Reliability Block Diagrams. Reliability block diagrams model the functioning of a complex system through use of a series of "blocks," in which each block represents the working of a system component or subsystem. Reliability block diagrams allow one to aggregate from component reliabilities to system reliability. A reliability block diagram can be used to optimize the allocation of reliability to system components by considering the possible improvement of reliability and the associated costs due to various design modifications. It is typical for very complex systems to initiate such diagrams at a relatively high level, providing more detail for subsystems and components as needed.

Fault Tree Analysis. Fault tree analysis is a systematic method for defining and analyzing system failures as a function of the failures of various combinations of components and subsystems. The basic elements of a fault tree diagram are events that correspond to improper functioning of components and subcomponents, and gates that represent and/or conditions. As is the case for reliability block diagrams, fault trees are initially built at a relatively coarse level and then expanded as needed to provide greater detail. The construction concludes with the assignment of reliabilities to the functioning of the components and subcomponents. At the design stage, these reliabilities can either come from the reliabilities of similar components for related systems, from supplier data, or from expert judgment. Once these detailed reliabilities are generated, the fault tree diagram provides a method for assessing the probabilities that higher aggregates fail, which in turn can be used to assess failure probabilities for the full system. Fault trees can clarify the dependence of a design on a given component, thereby prioritizing the need for added redundancy or some other design modification of various components, if system reliability is deficient. Fault trees can also assist with root-cause analyses.

Similarity Analysis

The two methods discussed above are "bottom-up" predictions. They use failure data at the component level to assign rates or probabilities of failure. Once the components and external events are understood, a system model is developed. An alternative method is to use a "top-down" approach using similarity analysis. Such an analysis compares two designs: a recent vintage product with proven reliability and a new design with unknown reliability. If the two products are very similar, then the new design is believed to have reliability similar to the predecessor design. Sources of reliability and failure data include supplier data, internal manufacturing test results from various phases of production, and field failure data.

There has been some research on similarity analyses, describing either the full process or specific aspects of this technique (see, e.g., Foucher et al., 2002). Similarity analyses have been reported to have a high degree of accuracy in commercial avionics (see Boydston and Lewis, 2009). Because this is a relatively new technique for prediction, however, there is no universally accepted procedure.

The main idea in this approach is that all the analysts agree to draw as much relevant information as possible from tests and field data. As the "new" product is produced and used in the field, these data are used to update the prediction for future production of the same product (for details, see Pecht, 2009). However, changes between the older and newer product do occur, and can involve

- product function and complexity
- technology upgrades
- engineering design process
- design tools and rules
- engineering team
- de-rating concepts
- assembly suppliers
- manufacturing processes
- manufacturing tooling
- assembly personnel
- test equipment and processes
- management policies
- quality and training programs, and
- application and use environment.

In this process, every aspect of the product design, the design process, the manufacturing process, corporate management philosophy, and quality processes and environment can be a basis for comparison of differences. As the extent and degree of difference increases, the reliability differences will also increase. Details on performing similarity analyses can be found in the *Guide for Selecting and Using Reliability Predictions* of the IEEE Standards Association (IEEE 1413.1).

REDUNDANCY, RISK ASSESSMENT, AND PROGNOSTICS

Redundancy

Redundancy exists when one or more of the parts of a system can fail and the system can still function with the parts that remain operational. Two common types of redundancy are active and standby.

In active redundancy, all of a system's parts are energized during the operation of a system. In active redundancy, the parts will consume life at the same rate as the individual components. An active redundant system is a standard "parallel" system, which only fails when all components have failed.

In standby redundancy, some parts are not energized during the operation of the system; they get switched on only when there are failures in the active parts. In a system with standby redundancy, ideally the parts will last longer than the parts in a system with active redundancy. A standby system consists of an active unit or subsystem and one or more inactive units, which become active in the event of a failure of the functioning unit. The failures of active units are signaled by a sensing subsystem, and the standby unit is brought to action by a switching subsystem.

There are three conceptual types of standby redundancy: cold, warm, and hot. In cold standby, the secondary part(s) is completely shut down until needed. This type of redundancy lowers the number of hours that the part is active and does not consume any useful life, but the transient stresses on the part(s) during switching may be high. This transient stress can cause faster consumption of life during switching. In warm standby, the secondary part(s) is usually active but is idling or unloaded. In hot standby, the secondary part(s) forms an active parallel system. The life of the hot standby part(s) is consumed at the same rate as active parts. Redundancy can often be addressed at various levels of the system architecture.

Risk Assessment

"Risk" is defined as a measure of the priority assessed for the occurrence of an unfavorable event. General methodologies for risk assessment (both quantitative and qualitative) have been developed and are widely available. The process for assessing the risks associated with accepting a part for use in a specific application involves a multistep process:

1. Start with a risk pool, which is the list of all known risks, along with knowledge of how those risks are quantified (if applicable) and possibly mitigated.
2. Determine an application-specific risk catalog: Using the specific application's properties, select risks from the risk pool to form an application-specific risk catalog. The application properties most likely to be used to create the risk catalog include functionality, life-cycle environments (e.g., manufacturing, shipping and handling, storage, operation, and possibly end-of-life), manufacturing characteristics (e.g., schedule, quantity, location, and suppliers), sustainment plans and requirements, and operational life requirements.

3. Characterize the risk catalog: Generate application-specific details about the likelihood of occurrence, consequences of occurrence, and acceptable mitigation approaches for each of the risks in the risk catalog.
4. Classify risks: Classify each risk in the risk catalog in one of two categories: functionality risks and producibility risks. Functionality risks impair the system's ability to operate to the customer's specification. They are risks for which the consequences of occurrence are loss of equipment, mission, or life. Producibility risks are risks for which the consequences of occurrence are financial (reduction in profitability). Producibility risks determine the probability of successfully manufacturing the product, which in turn refers to meeting some combination of economics, schedule, manufacturing yield, and quantity targets.
5. Determine risk-mitigating factors: Factors may exist that modify the applicable mitigation approach for a particular part, product, or system. These factors include the type or technology of the part under consideration, the quantity and type of manufacturer's data available for the part, the quality and reliability monitors employed by the part manufacturer, and the comprehensiveness of production screening at the assembly level.
6. Rank and down-select: Not all functionality risks require mitigation. If the likelihood or consequences of occurrence are low, then the risk may not need to be addressed. The ranking may be performed using a scoring algorithm that couples likelihood and consequence into a single dimensionless quantity that allows diverse risks to be compared. Once the risks are ranked, those that fall below some threshold in the rankings can be omitted.
7. Determine the verification approach: For the risks that are ranked above the threshold determined in the previous activity, consider the mitigation approaches defined in the risk catalog. The acceptable combination of mitigation approaches becomes the required verification approach.
8. Determine the impact of unmanaged risk: Combine the likelihood of risk occurrence with the consequences of occurrence to predict the resources associated with risks that the product development team chooses not to manage proactively. (This assumes that all unmanaged risks are producer risks.)
9. Determine the resources required to manage the risk: Create a management plan and estimate the resources needed to perform a prescribed regimen of monitoring the part's field performance, the vendor, and assembly/manufacturability as applicable.

10. Determine the risk impact: Assess the impact of functionality risks by estimating the resources necessary to develop and perform the worst-case verification activity allocated over the entire product life-cycle (production and sustainment). The value of the product that may be scrapped during the verification testing should be included in the impact. For managed producibility risks, the resources required are used to estimate the impact. For unmanaged producibility risks, the resources predicted in the impact analysis are translated into costs.
11. Decide whether the risk is acceptable: If the impact fits within the overall product's risk threshold and budget, then the part selection can be made with the chosen verification activity (if any). Otherwise, design changes or alternative parts must be considered.

Prognostics

A product's health is the extent of degradation or deviation from its "normal" operating state. Health monitoring is the method of measuring and recording a product's health in its life-cycle environment. Prognostics is the prediction of the future state of health of a system on the basis of current and historical health conditions as well as historical operating and environmental conditions.

Prognostics and health management consists of technologies and methods to assess the reliability of a system in its actual life-cycle conditions to determine the likelihood of failure and to mitigate system risk: for examples and further details, see Jaai and Pecht (2010) and Cheng et al. (2010a, 2010b). The application areas of this approach include civil and mechanical structures, machine-tools, vehicles, space applications, electronics, computers, and even human health.

Prognostics and health management techniques combine sensing, recording, and interpretation of environmental, operational, and performance-related parameters to indicate a system's health. Sensing, feature extraction, diagnostics, and prognostics are key elements. The data to be collected to monitor a system's health are used to determine the sensor type and location in a monitored system, as well as the methods of collecting and storing the measurements. Feature extraction is used to analyze the measurements and extract the health indicators that characterize the system degradation trend. With a good feature, one can determine whether the system is deviating from its nominal condition: for examples, see Kumar et al. (2012) and Sotiris et al. (2010). Diagnostics are used to isolate and identify the failing subsystems/ components in a system, and prognostics carry out the estimation of remaining useful life of the systems, subsystems, or components: for examples of diagnostics and prognostics, see Vasan et al. (2012) and Sun et al. (2012).

The prognostics and health management process does not predict reliability but rather provides a reliability assessment based on in-situ monitoring of certain environmental or performance parameters. This process combines the strengths of the physics-of-failure approach with live monitoring of the environment and operational loading conditions.

6

Reliability Growth Through Testing

There are two ways to produce a reliable system. One can design in reliability, and one can improve the initial design through testing. Chapter 5 discussed designing reliable systems; this chapter describes improving system reliability through testing. Because it is difficult to test long enough to experience a large number of failures, testing is often accelerated both to understand where reliability problems might surface and to assess system reliability. Given that, a large fraction of this chapter deals with accelerated testing and related ideas.

Reliability testing is used to identify failure modes and to assess how close a system is to the required reliability level. Reliability assessment is also important for understanding the capabilities and limitations of a system in operational use. Reliability testing (and assessment) can be divided into two separate issues. First, there is testing for the reliability of the system as produced (for instance, at the time of system acceptance). One might refer to this as out-of-the-box reliability. Second, there is testing for the reliability of the system after it has been in use for some time, that is, testing to predict long-term reliability performance.

BASIC CONCEPTS AND ISSUES

It is important to keep in mind that reliability is always a function of the environment and nature of use. Therefore, a reliability assessment needs to be a function of both the history of environments of use and of profiles of use (e.g., speed, payload, etc.). Also, when used for prediction, reliability assessments rely on the validity of the estimation models used in conjunction with the test data collected. The types of testing and estimation procedures preferred for use depend on the stage of development of the system and the purpose of the test.

We note that the systems referred to in this chapter are generic, encompassing full systems, subsystems, and components. Some of the testing techniques are only applicable for hardware systems (such as accelerated life testing), although other techniques described are applicable to both hardware and software systems, such as demonstration testing. As discussed in Chapter 3, we remind readers that different reliability metrics are appropriate for different kinds of systems.

Data from reliability tests are used to estimate current reliability levels through use of a (properly) selected reliability metric. One can use these assessed reliability levels to track the extent to which they approach the required level as the system improves. Tracking growth in reliability over time is important for discriminating between systems that are and are not likely to achieve their reliability requirements on the basis of their current general design scheme. By identifying systems that are unlikely to achieve their required reliability early in development, increased emphasis can be placed on finding an alternate system design, which might include using higher reliability parts or materials, or allocating additional reliability testing resources to identify additional sources of reliability failures.

Because it is extremely inefficient to make substantial design changes in later stages of development or after deployment, it is important to identify any problems in design or, more generally, the likely inability for a system to meet its reliability requirements, during the design and early development stages. Therefore, careful reliability testing of systems, subsystems, and components while the design remains in flux is crucial to achieving desired reliability levels in a cost-efficient way prior to fielding. During development, the program of test, analyze, fix, and test[1] can be used to identify and eliminate design weaknesses inherent to intermediate prototypes of complex systems. Using this approach is generally referred to as "reliability growth." Specifically, reliability growth is the improvement in the true but unknown initial reliability of a developmental program as a result of failure mode discovery, analysis, and effective correction.

In addition to testing during development, feedback from field use or from tests after fielding can also be used to improve system design and, consequently, the system's long-term reliability. However, as discussed in previous chapters, postdevelopment redesign is very cost inefficient in comparison with finding reliability problems earlier during the design and

[1]Note the difference between this approach and that of test, analyze, and fix, discussed in Chapter 4: this approach adds a second test that can be useful in providing an assessment of the success of the fix.

development stages. Yet, it is still useful to point out that corrective actions can extend beyond initial operational test and evaluation. One or more focused follow-on tests and evaluations can be conducted after the initial operational test and evaluation, allowing previously observed deficiencies and newly implemented redesigns or fixes to be examined.

RELIABILITY TESTING FOR GROWTH AND ASSESSMENT

Several kinds of reliability tests are typically used in industry. Some of them are useful for identifying undiscovered failure modes, and some of them are useful for estimating current reliability levels. In this section we discuss three of them: highly accelerated life testing; reliability demonstration or acceptance tests; and accelerated life testing and accelerated degradation testing.

Highly Accelerated Life Testing

Highly accelerated life testing (HALT) is an upstream method of discovering failure modes and design weaknesses. HALT tests use extreme stress conditions to determine the operational limits of systems, which are the limits beyond which various failure mechanisms will occur other than those that would occur under typical operating conditions. HALT is primarily used during the design phase of a system.

In a typical HALT test, the system (or component) is subject to increasing levels of temperature and vibration (independently and in combination) as well as rapid thermal transitions (cycles) and other stresses related to the intended environments of use of the system. In electronics, for example, HALT can be used to locate the causes of the malfunctions of an electronic board. These tests can also include extreme humidity or other moisture, but because the effect of humidity on a system's reliability requires a long time to assess, HALT is typically conducted only under the two main stresses of temperature and vibration. The results of HALT tests enable the designer to make early decisions regarding the components to be used in the system.

The results from HALT tests are *not* intended for reliability assessment because of the short test periods and the extreme stress levels used. Indeed, HALT is not even a form of accelerated life testing (see below) because its focus is on testing the product to induce failures that are unlikely to occur under normal operating conditions.[2] One goal of HALT is to determine the root cause of potential failures (see Hobbs, 2000). The stress range and methods of its application in HALT (e.g., cyclic, constant, step increases)

[2]In accelerated life testing, the linkage between accelerated use and normal use is modeled to provide reliability assessments.

are dependent on the component to be tested, its requirements and failure modes, and the stresses to which it will be subject. Because such knowledge may only be held by the developer, it is important that such testing be conducted prior to delivery to the U.S. Department of Defense (DoD). Given that DoD can use such information to help design its own developmental tests, it is important that records of such testing, including the stresses applied, the failures discovered, and any design modifications taken in response, be made available to DoD to guide its testing.

Reliability Demonstration or Production Reliability Acceptance Test

When a milestone of a system in development is completed, in order to ascertain whether reliability growth of a component or subsystem is satisfactory, the contractor can carry out a reliability demonstration test. The basic idea is that, under normal operating conditions, a number of units are tested for a specified amount of time, and the resulting data are used to assess whether the observed reliability is reflective of a target value. Given the necessary number of units under test to provide such information, such tests are generally done at the component or subsystem level, rather than at the full system level.

The goal of a production reliability acceptance test is similar to that of the reliability demonstration test, but it is undertaken when a contractor intends to deliver a batch of products for actual use or inventory, and the contractor and DoD have to design a test plan that ensures that the probability of accepting a batch of production that has defective products and the probability of rejecting a good batch are both small.

Accelerated Life Testing and Accelerated Degradation Testing

In many cases, accelerated life testing (ALT) may be the only viable approach to assess whether a component or subsystem can be expected to meet a requirement for reliability over its lifetime, in contrast to the reliability prior to initial use. ALT can be conducted using three different approaches, one for testing full systems and two others that are more relevant for tests of subsystems and components. The first approach is conducted by accelerating the "use" of the system at normal operating conditions, such as in the case of systems that are used only a fraction of the time in a typical day. Examples include home appliances and auto tires, which are tested under use for, say, 24 hours, rather than for the much shorter period of time that would usually be used.

The second approach is generally carried out for components and subsystems relatively early in system development by subjecting a sample of components or subsystems to stresses that are more severe than normal operating conditions. The third approach is referred to as accelerated degradation testing: it is used to examine systems for signs of degradation rather than outright failure. It is conducted by subjecting components or subsystems that exhibit some type of degradation such as stiffness of springs, corrosions of metals, and wear-out of mechanical components to accelerated stresses. Both accelerated life tests and accelerated degradation tests are most useful in situations in which there is a predominant failure mode that is a function of a single type of stress. Consequently, these techniques are commonly applied at the component level, rather than the full-system level.

The reliability data obtained from ALT are used to estimate the parameters of a model that predicts the reliability of the component, subsystem, or system under normal operating conditions. This model is either statistics or physics based, and it is used to link the failure time distribution for time under normal use to a failure time distribution for time under extreme use. The assessed validity of such models should affect the degree to which the resulting estimates are trusted, which in turn could affect decisions about system redesign and determination of preventive maintenance schedules. For example, if the reliability prediction based on ALT results shows that the units exhibit constant failure rates, then it is not reasonable to conduct preventive maintenance for such units, because older units are not less reliable than new units. In contrast, if the units exhibit increasing failure rates (e.g., from wear-out), then plant maintenance or condition-based maintenance strategies would be economical to implement (for details, see Elsayed, 2012).

Reliability estimates from ALT depend not only on the linkage models, but also on the experimental design of the test plans. Stress loadings, such as constant stress, ramp stress, or cyclic stress; the allocation of test units to stress levels; the number of stress levels; the appropriate test duration; and other experimental variables can improve the accuracy of the resulting reliability estimates.

It may be that initial guesses at a model to link extreme to normal stresses may turn out to be invalid for many ALT situations. Therefore, to better assess long-run reliability, it would be useful for DoD and contractors to work together closely to determine both good designs of accelerated life tests and acceptable reliability prediction models based on subject-matter assumptions that are agreed to be reasonable.

Most reliability data from ALT are time-to-failure measurements obtained from testing samples of units at different stresses and noting failures. However, particularly for tests at stress levels close to normal operating conditions, instead of failing, components may suffer measurable degradation as a prelude to failure. For example, a component may start a test with an acceptable resistance value, but as test time progresses the resistance may drift so that it eventually reaches an unacceptable level that causes the component to fail.

In such cases, measurements of the degradation of the characteristics of interest (those whose degradation will ultimately result in failure of the part) are frequently taken during the test. The degradation data are then analyzed and used to predict the time to failure at normal conditions. We refer to this as accelerated degradation testing, which requires a reliability prediction model to relate degradation results of a test under accelerated conditions to failures under normal operating conditions. Proper identification of the degradation indicator is critical for the analysis of degradation data and subsequent decisions about maintenance schedules and replacements. An example of such an indicator is hardness, which is a measure of degradation of elastomers. Other indicators include loss of stiffness of springs, corrosion rate of beams and pipes, and crack growth in rotating machinery. In some cases, the degradation indicator might not be directly observed, and destruction of the unit under test is the only alternative available to assess its degradation. This type of testing is referred to as accelerated destructive degradation testing.

In some applications, it is possible to use accelerated degradation testing instead of accelerated life testing. Degradation tests often provide more information for the same number of test units, as well as information that is more directly related to the underlying failure mechanism, which often provides a sounder basis for determining a model that can be used for extrapolation to use conditions. This advantage comes at the cost of needing to validate the model linking the current degree of degradation and the distribution of remaining system lifetime. This might be done by using the degradation data to effectively predict the time when the degradation of the unit crosses a specified threshold level. Therefore, if feasible, and if there is a well-defined degradation model, an accelerated degradation test would be an effective approach for predicting system reliability after different amounts of use. It is important to note, however, that some systems do not exhibit degradation during use before the occurrence of sudden failures.

The design of ALT has experienced many advances over the past decades; Elsayed (2012) provides a description of many of the recent ideas. They include designs to measure the following stress types: mechanical stresses, which are often a result of fatigue (due to elevated temperature, shock and vibration, and wear-out); electrical stresses (e.g., power cycling, electric field, current density, and electromigration); and environmental stresses (e.g., humidity, corrosion, ultraviolet light, sulfur dioxide, salt and fine particles, alpha rays, and high levels of ionizing). There are different ways stress can be applied, including constant level, step-stresses (either low-to-high or high-to-low), cyclic loading, power-on power-off, ramp tests, and various combinations of these. Elsayed (2012, p. 7) notes: "[D]ue to tight budgets and time constraints, there is an increasing need to determine the best stress loading in order to shorten the test duration and reduce the total cost while achieving an accurate reliability prediction." There is now a growing literature, cited in Elsayed (2012) on ALT designs that indicate what types of designs are better for different types of components and subsystems.

Given the important benefits from effective testing for reliability growth and for reliability assessment, the panel recommends that DoD take several steps to ensure that contractors use tests that are capable of measuring agreed-on metrics; that the designs of test plans and the validation of reliability models linking extreme to normal use and degradation to failures are examined by reliability engineers prior to application; and that contractors supply DoD with all test data relevant to reliability assessment (see Recommendation 12 in Chapter 10).[3] We also recommend creation of a database that includes the results not only from contractor testing, but also from DoD developmental and operational testing and from field use. Such a database would support the model validation for accelerated life testing and accelerated degradation testing. In addition, for degradation testing, measurement of the degree of degradation should also be collected as part of this database. In addition, such databases could also include the estimated reliability performance of fielded systems to provide better "true" values for reliability attainment. Finally, there is also a need to save sufficient detail describing the fielding environment(s), including the technology type and the specified design temperature limits.

[3]All the panel's recommendations are presented in Chapter 10.

7

Developmental Test and Evaluation

This chapter describes the role that developmental testing plays in assessing system reliability. The requirements for a system specify the functions it is expected to carry out and the operational situations in which it is expected to do so. The goal of testing—particularly in situations where theory or prior experience do not predict how well a system will function in specific environments—is to show whether or not the system will function satisfactorily over the specified operational conditions. Thus, the goal of the design of developmental testing is to be able to evaluate whether a system can do so. In our context, it is to assess whether it will be reliable when deployed.

Developmental testing, like all forms of testing, is not a cost-effective substitute for thorough system and reliability engineering, as a system is developed from concept to reality. However, developmental testing is an essential supplement. Testing can provide the hard, empirical evidence that the system works as designed or that there are reliability problems that the system designers did not anticipate.

For complex systems intended for use in multidimensional environments, designing an efficient, effective, and affordable testing program is difficult. It requires a mix of system engineering, subject-matter knowledge, and statistical expertise. For the U.S. Department of Defense (DoD), there are a wide variety of defense systems, and there is no "one-size-fits-all" menu or checklist that will assure a satisfactory developmental test program. The knowledge, experience, and attitudes of the people involved will be as important as the particular methods that are used. Those methods will have to be tailored to particular situations: hence, there is a need for the people involved to have the requisite knowledge in reliability engineering and statistics to adapt methods to the specific systems.

In this chapter, the first two sections look briefly at the role of contractors in developmental testing and basic elements of developmental testing both for contractors and for DoD. We then describe in more detail three aspects of developmental testing: experimental design to best identify reliability defects and failure modes and for reliability assessment; data analysis of reliability test results; and reliability growth monitoring, which has as its goal identification of when a system is ready for operational testing.

CONTRACTOR TESTING

We recommend (see Recommendation 12 in Chapter 10) that contractors provide DoD with detailed information on all tests they perform and the resulting data from those tests. If this recommendation is adopted, then early developmental testing becomes a continuation and extension of contractor testing, which focuses primarily on identifying defects and failure modes, including design problems and material weaknesses. However, as the time for developmental testing approaches, contractors likely carry out some full-system testing in more operationally relevant environments: this testing is similar to the full-system testing that will be carried out at the end of developmental testing by DoD. Making use of more operationally relevant full-system testing has the important benefit of reducing surprises and assuring readiness for promotion to developmental testing. Given the potential similarity in the structure of the tests, this contractor testing would also increase the opportunities for combining results from contractor and later developmental (and operational) DoD tests.

While early developmental testing emphasizes the identification of failure modes and other design defects, later developmental testing gives greater emphasis to evaluating whether and when a system is ready for operational testing.

BASIC ELEMENTS OF DEVELOPMENTAL TESTING

Several elements are important to the design and evaluation of effective developmental tests: statistical design of experiments, accelerated tests, reliability tests, testing at various levels of aggregation, and data analysis:

- Statistical design of experiments involves the careful selection of a suite of test events to efficiently evaluate the effects of design and operational variables on component, subsystem, and system reliability.

- Accelerated tests include accelerated life tests and accelerated degradation tests, as well as highly accelerated life tests, which are used to expose design flaws (see Chapter 6).
- Reliability tests, which are often carried out at the subsystem level, include tests to estimate failure frequencies for one-shot devices, mean time to failure for continuously operating, nonrepairable devices, mean time between failures for repairable systems, and probabilities of mission success as a function of reliability performance for all systems.
- Testing at various levels of aggregation involves testing at the component, subsystem, and system levels. Generally, more lower-tiered testing will be done by the contractor, while at least some full-system testing is best done by both the contractor and DoD.
- Analysis of developmental test data has two goals: (1) tracking the current level of reliability performance and (2) forecasting reliability, including when the reliability requirement will be met. If contractor and developmental test environments and scenarios are sufficiently similar, then combining information models, as described in National Research Council (2004), should be possible. However, merging data collected under different environmental or stress conditions is very complicated, and attempts to do so need to explicitly account for the different test conditions in the model.[1]

To make the best use of developmental testing, coordination between the contractor and government testers is important. Effective coordination requires a shared view that testing should be mutually beneficial for the contractor, DoD, and taxpayers in achieving reliability growth. Testing from an adversarial perspective does not serve any of the parties or the ultimate system users well.

If, as we recommend, contractor test data are shared with the DoD program personnel, then those data can provide a sound basis for subsequent collaboration as further developmental testing is done in order to improve system reliability, if needed, and to demonstrate readiness for operational testing by DoD. There are also technical aspects to the recommended collaboration, such as having DoD developmental test design reflect subject-matter knowledge of both the developer and the user. This collaboration includes having a test provide information for what are believed to be the

[1]This is what is done when developing models to link accelerated test results to inference for typical use, but the idea is much more general. It includes accounting for the developmental test/operational test (DT/OT) gap in reliability, as in Steffey et al. (2000), and providing estimates for untested environments that are interpolations between environments in which tests were carried out.

most important design and operational variables and the most relevant levels of those variables, agreement on the reliability metrics, and, more broadly, defining a successful test.

One example of collaboration stems from the need for reliability performance to be checked across and documented for the complete spectrum of potential operating conditions of a system. If an important subset of the space of operational environments was left unexplored during contractor testing—such as testing in cold, windy, environments—it would be important to give priority during developmental testing to the inclusion of test replications in those environments (see Recommendation 11 in Chapter 10).

DESIGNED EXPERIMENTS

Developmental tests for reliability are experiments. Their purpose is to evaluate the impact of operational conditions on system reliability. To be efficient and informative, it is critical to use the principles of statistical experimental design to get the most information out of each test event. A recent initiative of the Director of Operational Test and Evaluation (DOT&E, October, 2010, p. 2) is summarized in a memorandum titled "Guidance on the Use of Design of Experiments (DOE) in Operational Test and Evaluation," and very similar guidance applies to developmental testing:

1. A clear statement of the goal of the experiment, which is either how the test will contribute to an evaluation of end-to-end mission effectiveness in an operationally realistic environment, or how the test is likely to find various kinds of system defects.
2. The mission-oriented response variables for (effectiveness and) suitability that will be collected during the test (here, the reliability metrics).
3. Factors that (may) affect these response variables. Test plans should provide good breadth of coverage of these factors across their applicable levels. Further, test plans should select test events that focus on combinations of factors of greatest interest, and test plans should select levels of these factors that are consistent with the test goals.
4. The appropriate test matrix (suite of tests) should be laid out for evaluating the effects of the controlled factors in the experiment on the response variables.[2]

[2]We note that specific designs that might prove valuable in this context, given the large number of possible factors affecting reliability metrics, include fractional factorial designs, which maximize the number of factors that can be simultaneously examined, and Plackett-Burman designs, which screen factors for their importance: for details, see Box et al. (2005).

5. Determination of the experimental units and blocking.
6. Procedures that dictate control of experimental equipment, instrumentation, treatment assignments, etc. should be documented for review.
7. Sufficient test scenario replications should be planned to be able to detect important effects and estimate reliability parameters.

The panel strongly supports this guidance. The main additional issue that we raise here is the degree of operational realism that is used in the non-accelerated developmental tests. Using operationally relevant environments and mission lengths is important for both identifying defects and for evaluation. It is well known that system reliability is often assessed to be much higher in developmental tests than in operational tests (see Chapter 8), which is referred to as the DT/OT gap. Clearly, some failure modes appear more frequently under operationally relevant testing. Therefore, to reduce the number of failure modes left to be discovered during operational testing, and at the same time have a better estimate of system reliability in operationally relevant environments, non-accelerated developmental tests should, to the extent possible, subject components, subsystems, and the full system to the same stresses that would be experienced in the field under typical use environments and conditions. This approach will narrow the potential DT/OT gap in reliability assessments and provide an evaluation of system reliability that is more operationally relevant.

TEST DATA ANALYSIS

A primary goal of reliability testing is to find out as much as possible about what conditions of use contribute to the system being more or less reliable. This goal then supports a root-cause analysis to determine why those conditions caused those reductions in reliability. Therefore, the object of most developmental test data analysis is to measure system reliability as a function of the factors underlying the test environments and the missions. To do so, it is necessary to distinguish between actual increases and decreases in system reliability and natural (within and between) system variation.[3]

Given the limited number of replications of reliability tests and therefore limited ability to identify differences in reliability between scenarios of use, as well as the high priority of determining when requirements have

[3]We define within-system variation as the variability in the performance of a given system in a given environment over replications and between-system variation as the variability in performance in a given environment between different prototypes produced using the same design and manufacturing process.

been met and a system can be approved for operational testing, it is not surprising that learning about differences in performance between scenarios of use is sometimes ignored. The reliability requirement to be assessed is typically an average reliability taken over operationally relevant environments of use: that requirement is often taken from the operational mode summary/mission profile (OMS/MP). This average is compared with the requirement, which is similarly defined as the same type of average. Although interest in this average is understandable, it is also important to have assessments for the individual environments and missions, to the extent that that it is feasible given test budgets.

Another important type of disaggregation is distinguishing between two different ways of treating failure data from reliability tests. One could look at any failure as a system failure and make evaluations and assessments based on the total number of failures that occur. However, the process that generates some types of failures may be quite different from the processes generating others, and there may be advantages (statistical and operational) to analyzing different types of failure modes separately. For example, reliability growth modeling may produce better estimates if failure modes are categorized into hardware and software failures for analytic purposes and then such estimates are aggregated over failure type to assess system performance.

Systems themselves can be usefully grouped into three basic types with respect to reliability performance (see Chapter 3), and the preferred analysis to be carried out depends on which type of system is under test. The basic types are one-shot devices, continuously operating systems that are not repairable, and continuously operating system that are repairable.

One-Shot Systems

For one-shot systems, the primary goal of developmental test data analysis is to estimate the average failure probability. However, it is also important to communicate to decision makers the imprecision of the average estimated failure probabilities, which can be done through the use of confidence intervals. As mentioned above, to the extent possible, given the number of replications, it would also be useful to provide estimated probabilities and confidence intervals disaggregated by variables defining mission type or test environment.

In doing so, it is important not to pool nonhomogeneous results. For example, if the test results indicate high failure probabilities for high temperatures but low failure probabilities at ambient temperatures (based on enough data to detect important differences in underlying failure probabilities), then one should report separate estimates for these two different types of experimental conditions, rather than pool them together for a combined estimate. (Of course, given that the requirement is likely to be such a pooled estimated, its estimate must be provided.)

The common practice of reporting only pooled failure data across multiple mission profiles or environments (e.g., high and low temperature test results) does not serve decision makers well. Discovering and understanding differences in failure probabilities as a function of mission variables or test environments is important for correcting defects. If such defects cannot be corrected, then it might be necessary to redefine the types of missions for which the system can be used.

Nonrepairable Continuously Operating Systems

For nonrepairable continuously operating systems, the goal of the developmental test data analysis is to estimate the lifetime distribution for the system, to the extent possible, as a function of mission design factors. Such an estimate would be computed from lifetime test data. In planning such a test, it would be best to run the test at least long enough to cover mission times of interest and with enough test units to provide sufficient precision (as might be quantified by the width of a confidence interval).

To understand the dependence on design factors, one would develop a statistical model of the lifetime distribution using the design factors as predictors and using the test data to estimate the parameters of such a model. Such a model may need to include estimates of the degree to which the reliability of the system was degraded by storage, transport, and similar factors. The fitted model can then be used to estimate the probability of the system's working for a period of time greater than or equal to the mission requirement, for the various types of missions the system will be used for. For these tests, too, it is important to provide information on the uncertainty of the estimates, usually expressed through use of confidence intervals. In some cases, resampling techniques may be needed to produce such confidence intervals.

It is common for DoD to use mean time to failure as a summary metric to define system reliability requirements for continuously or intermittently operating systems. Although mean time to failure is a suitable metric for certain components that are expected to wear out and be replaced, such as a battery, it is inappropriate for high-reliability components, such as integrated circuit chips, for which the probability of failure is small over the technological life of a system. In the latter case, a failure probability or quantile in the lower tail of the distribution would be better. In addition, given missions of a specific duration, it is important to measure the distribution of time to failure, from which one can estimate the probability of mission success, not necessarily under the assumption of an exponential distribution of time to failure.

Repairable Continuously Operating Systems

For repairable systems, the mean time between failures is a reasonable metric when failures in time can be assumed to be described by a Poisson process. However, if the underlying failure mechanism is governed by a nonhomogeneous Poisson process (such as the AMSAA-CROW model[4]) that has a nonconstant rate of occurrence of failures, mean time between failures would be a misleading metric. In such cases, one should instead study the average cumulative number of failures up to a given time. Ideally, a parametric formulation of the nonconstant rate of occurrence of failures is used, and reliability is assessed through the parameter estimates. A step-intensity or piecewise-exponential model can be used for reliability growth data that are collected from a developmental test in order to emphasize the effect of the design changes.

Merging Data

Because the time on test for any individual prototype and for any design configuration is often insufficient to provide high-quality estimates of system reliability, methods that attempt to use data external to the tests to augment developmental test data are worth considering. Several kinds of data merging are possible: (1) combining test results across tests of the system for different levels of aggregation, (2) combining information from different developmental tests either for the same system or related systems, and (3) combining developmental and contractor test data, although this raises issues of independence of government assessment.

In some cases, one will have useful data from testing or other information at multiple system levels: that is, one will often have data not only on full system reliability, but also on component and subsystem reliabilities. In those cases, one may be able to use models, such as those implied by reliability block diagrams, to produce more precise estimates of system reliability through merging this information. Of course, there is always concern about combining information from disparate experimental conditions. However, if such differences can be handled by making adjustments, then one might be able to produce system reliability estimates with associated confidence limits based on an ensemble of multilevel data, for which the estimates would be preferred to the estimates using only the test data for tests on the full system. The primary work in this area is the PREDICT methodology developed at Los Alamos (see, e.g., Wilson et al., 2006; Hamada et al., 2008; Reese et al., 2011). The development

[4]This is a reliability growth model developed by Crow (1974) and first used by the U.S. Army Materiel Systems Analysis Activity: see Chapter 4.

of such models will be situation specific and may not provide a benefit in some circumstances.

In some situations there is relevant information from tests on previous versions of the same system or similar systems or on systems with similar components or subsystems. In such situations, the assumptions about a prior distribution may be clearly seen to be valid, so that Bayesian methods for combining information may be able to produce preferred estimates to those based only on the data from tests on the current system. An example would be information about the Weibull shape parameter based on knowledge of an individual failure mode in similar systems (see, e.g., Li and Meeker, 2014).

In using Bayesian techniques, it is crucial to document exactly how the prior distributions were produced, and the validation of the use of such priors, and to assess the sensitivity of the estimates and measures of uncertainty as a function of the specification of the prior distribution. There are also non-Bayesian methods for combining data, such as the Maximus method for series and parallel systems and extensions of this research, although some aspects of inference may be somewhat more complicated in such a framework (see, e.g., Spencer and Easterling, 1986). In general, it is reasonable to assume that combining information models would be more useful for one-shot and nonrepairable systems than for repairable systems.

One way of combining data across tests of a system as the system undergoes changes to address discovered defects and failure modes is to use reliability growth modeling (see Chapter 4). However, such models cannot accommodate tests on different levels of the system and cannot use information on the environments or missions under test.

Finally, combining information over developmental tests is complicated by the fact that design defects and failure modes discovered during developmental testing often result in changes to the system design. Therefore, one is often trying to account not only for differences in the test environment, but also for the differences in the system under test. We recommend (Recommendation 19, in Chapter 10) that the delivery of prototypes to DoD not occur until a system's performance is assessed as being consistent with meeting the requirement. If that recommendation is adopted, then it would limit the number of defects needing to be discovered to a small number, which would result in the systems in developmental testing undergoing less change, which would greatly facilitate the development of combining information models.

RELIABILITY GROWTH MONITORING

Monitoring progress toward meeting reliability requirements is now mandated by a directive-type memorandum DTM 11-003 (U.S. Department of Defense, 2013b, p. 3), which states

> Reliability Growth Curves (RGC) shall reflect the reliability growth strategy and be employed to plan, illustrate, and report reliability growth. A RGC shall be included in the SEP at MS A, and updated in the TEMP beginning at MS B. The RGC will be stated *in a series of intermediate goals and tracked through fully integrated, system-level test and evaluation events until the reliability threshold is achieved* [emphasis added]. If a single curve is not adequate to describe overall system reliability, curves will be provided for critical subsystems with rationale for their selection.

At least three technical issues need to be faced in satisfying this mandate. First, how are such intermediate goals to be determined? Chapter 4 presents a discussion of the value of formal reliability growth modeling when used for various purposes. As argued there, under certain assumptions, formal reliability growth models could at times produce useful targets for system reliability as a function of time in order to help discriminate between systems that are or are not likely to meet their reliability requirements before operational tests are scheduled to begin. Oversimplifying, one would input the times when developmental tests were scheduled into a model of anticipated reliability growth consistent with meeting the requirement just prior to operational testing and compare the observed reliability from each test with the model prediction for that time period.

Unfortunately, the most commonly used reliability growth models have deficiencies (as discussed in Chapter 4). Given the failure to represent test circumstances in the families of reliability growth models commonly used in defense acquisition, such models will often not provide useful estimates or predictions of system reliability under operationally relevant conditions. To address this deficiency, whenever possible, system-level reliability testing should be conducted under OMS/MP-like conditions, that is, under operationally realistic circumstances. Such models also assume monotonic improvement in reliability over time, but if there have been some major design changes, reliability might not be monotonically increasing until the changes are fully accommodated in terms of interfaces and other factors. Therefore, if such tests include a period or periods of development in which major new functionality was added to the system, then the assumption of monotonic reliability growth could possibly not hold, which could result in poor target values.

Since DTM 11-003 does not specify what models to use for this purpose, analysts are free to make use of alternative models that do take the specific test circumstances into consideration. We encourage efforts to develop reliability growth models that represent system reliability as a function of the conditions of the test environments, along the lines of physics-of-failure models (see Chapter 5).

Second, how should one produce current estimates of system reliability? It is likely that most developmental tests will be fairly short in duration and will rely on a relatively small number of test units because of the need to budget an unknown number of future developmental tests to evaluate future design modifications. As mentioned above, to supplement a limited developmental test in order to produce higher quality reliability estimates, assuming the tests are relatively similar, one could smooth the results of several test events over time, or fit some kind of parametric time series model, to model the growth in reliability. However, it is much more likely that the developmental tests will differ in important ways, which would reduce the applicability of such approaches. For instance, some of the later developmental tests may use more realistic test scenarios than the earlier ones, or different test scenarios.

More fundamentally, some tests are likely to use acceleration of some type, and some will not; moreover, some will be at the component or subsystem level, and some will be at the system level. Therefore, any type of combining information across such tests would be challenging, and would have to develop something similar to PREDICT to use the information from these various developmental tests. Of course, one could increase the duration and sample size of a developmental test so that no modeling was required to produce high-quality estimates, but this is unlikely to happen given current test resources.

Third, given that one is comparing model-based target values with current measures of system reliability, one would need to take into consideration the uncertainty of the current estimates of system reliability so that one can formulate decision rules with good type I and type II error rates. This consideration will ensure that such decision rules are formulated so that the systems that ultimately meet their reliability requirement are rarely flagged and systems that fail to meet their reliability requirement are frequently flagged. But developing confidence intervals for estimates based on merged test information may not be straightforward.

It is unclear how the uncertainty in estimated reliability is currently handled in DoD in analogous situations, such as determining whether a performance characteristic in an operational test is consistent with a requirement. In that application, one interpretation is that a confidence interval for the estimated reliability needs to lie entirely above the requirement in order for the system to be judged as satisfying the requirement. Given the substantial uncertainty in reliability assessments common in operational test evaluation, this would be an overly strict test that would often fail systems that had in fact met the requirement. (In other words, this rule would have a very high producer's risk.) Another possibility is that the confidence interval for the estimated reliability would only need to include the requirement to pass the operational test. This rule has a large consumer risk in small tests, because one could then have a system with a substantially lower reliability than the requirement, but if the uncertainty was large, one could not reject the hypothesis that the system had attained the requirement. In fact, with the use of such a decision rule, there would be an incentive to have small operational tests of suitability in order to have overly wide confidence intervals that would be likely to pass such systems.

A preferred approach would be for DoD to designate a reliability level that would be considered to be the lowest acceptable level, such that if the system were performing at that level, it would still represent a system worth acquiring. The determination of this lowest acceptable level would be done by the program executive office and would involve the usual considerations that make up an analysis of alternatives. Under this approach, a decision rule for proceeding to operational testing could be whether or not a lower confidence bound, chosen by considering both the costs of rejecting a suitable system and the costs of accepting an unsuitable system, was lower than this minimally acceptable level of reliability. Such a decision rule is, of course, an oversimplification, ignoring the external context, which might decide that an existing conflict required the system to be fielded even if it had not met the stated level of acceptability.

This technique could be easily adapted to monitoring reliability growth by comparing the estimated reliability levels to the targets produced using reliability growth models. As mentioned above, such a decision rule would also have to take into consideration any differences between test scenarios used to date and those used in operational testing, which is similar but somewhat more general than the problem we have been referring to as the DT/OT gap. Furthermore, this decision rule would also have to make some accommodation for model misspecification in the development of the target values, because the model used for that purpose would not be a perfect representation of how reliability would grow over time.

Finally, the last sentence quoted above from DTM 11-003 raises an additional technical issue, namely, when would it be useful to have reliability growth targets for subsystems as well as for a full system. The panel offers no comment on this issue.

8

Operational Test and Evaluation

Following our discussion of the role of developmental testing in assessing system reliability in Chapter 7, this chapter covers the role of operational testing in assessing system reliability. After a defense system has been developed, it is promoted to full-rate production and the field if it demonstrates in an operational test that it has met the requirements for its key performance parameters. The environment of operational testing is as close to field deployment as can be achieved, although the functioning of offensive weapons is simulated, and there are additional constraints related to safety, environmental, and related concerns.

In this chapter we first consider the timing and role of operational testing and then discuss test design and test data analysis. The final section considers the developmental test/operational test (DT/OT) gap.

TIMING AND ROLE OF OPERATIONAL TESTING

The assessment of the performance of defense systems, as noted in Chapter 1, is separated into two categories, effectiveness and suitability, a major component of the latter being system reliability. Operational testing is mainly focused on assessing system effectiveness, but it is also used to assess the reliability of the system as it is initially manufactured. Even though reliability (suitability) assessment is a somewhat lower priority than effectiveness, there is strong evidence (see below) that operational testing remains important for discovering reliability problems. In particular, the reliability estimates from operational testing tend to be considerably lower than reliability estimates from late-stage developmental testing. Although the benefit of operational testing is therefore clear for discovering unexpected reliability problems, because operational testing is of short duration it is not well-suited for identifying reliability problems that relate to longer use, such as material fatigue, environmental effects, and aging.

Moreover, although operational testing may identify many initial reliability problems, there is a significant cost for defense systems that begin operational testing before the great majority of reliability problems have been discovered and corrected. Design changes made during operational testing are very expensive and may introduce additional problems due to interface and interoperability issues. Given that many reliability problems will not surface unless a system is operated under the stresses associated with operational use, and given that operational testing should not be relied on to find design problems for the reasons given above, the clearly cost-efficient strategy would be to require that defense systems are not considered ready for entry into operational testing until the program manager for the U.S. Department of Defense (DoD), the contractor, and associated operational test agency staff are convinced that the great majority of reliability problems have been discovered and corrected. One way to operationalize this strategy is to require that defense systems not be promoted to operational testing until the estimated reliability of the system equals its reliability requirement under *operationally relevant conditions*. An important additional benefit of this strategy would be that it would eliminate, or greatly reduce, the DT/OT reliability gap (discussed below), which currently complicates the judgment as to whether a system's reliability as assessed in developmental testing is likely to be found acceptable in operational testing.

In addition to using developmental tests that include stresses more typical of real use (see Chapter 7), there are three additional steps that the DoD could adopt to reduce the frequency of reliability problems that first appear in operational testing.

First, DoD could require that contractor and developmental test interim requirements for reliability growth, as specified in the approved test and evaluation master plan, have been satisfied. Failure to satisfy this requirement, especially when substantial shortcomings are evident late in the developmental test program, is a strong indicator that the system is critically deficient.

Second, DoD could specify that reports (i.e., paper studies) of planned design improvements are, by themselves, inadequate to justify promotion to operational testing. Instead, DoD could require that system-level tests be done to assess the impact of the design improvements.

Third, DoD could require that reliability performance from contractor and developmental tests are addressed and documented for the complete spectrum of potential system operating conditions given in the operational mode summary/mission profile,[1] including both hardware and software performance. Doing so might help indicate areas that have not been fully tested under operationally relevant conditions. Separate software testing, possibly disconnected from hardware presence, also may be needed to fully explore a system's performance profile.

There are some other respects in which operational testing is limited, which places yet additional weight on developmental testing as the primary opportunity for discovery of design problems. Operational testing presents, at best, a snapshot of system reliability for new systems for two major reasons. First, the results are somewhat particular to the individual test articles. Second, the prototypes used in operational testing will be used only for short durations, not even close to the intended lifetimes of service or availability expected of the system. Thus, until developmental testing makes greater use of accelerated life testing in various respects (see Chapter 6), along with more operational realism in non-accelerated tests, there remains a good chance that some reliability problems will not appear until after a system is deployed. To address this problem, we recommend that DoD institute a procedure that requires that the field performance of post-operational test configurations (i.e., new designs, new manufacturing methods, new materials, or new suppliers) are tracked and the data used to inform additional acquisitions of the same system, and for planning and conducting future acquisition programs of related systems (see Recommendation 22 in Chapter 10).

Operational testing is limited to the exercising of the system in specific scenarios and circumstances (based, to a great extent, on the operational mode summary/mission profile). Given the limited testing undertaken in operational testing and the resulting limited knowledge about how the system will perform under other circumstances of operational use, due consideration should be taken when developing system utilization plans, maintenance, and logistics supportability concepts, and when prescribing operational reliability and broader operational suitability requirements.

Although this discussion has stressed the differences between developmental and operational testing, there are also strong similarities. Therefore, much of the discussion in Chapter 7 (and other chapters) about the importance of experimental design and data analysis for developmental testing also apply to operational testing. In the next two sections on design and data analysis we concentrate on issues that are particularly relevant to operational testing.

---

[1]This profile "defines the environment and stress levels the system is expected to encounter in the field. They include the overall length of the scenarios of use, the sequence of missions, and the maintenance opportunities" (National Research Council, 1998, p. 212).

TEST DESIGN

The primary government operational test event is the initial operational test and evaluation. It is an operationally realistic test for projected threat and usage scenarios, and it includes production-representative hardware and software system articles, certified trained operators and users, and logistics and maintenance support compatible with projected initial fielding.

Ideally the extent of the initial operational test and evaluation event would be of sufficient duration to provide a reasonable stand-alone assessment of initial operational reliability, that is, it should be of sufficient length to provide an acceptable level of consumer risk regarding the reliability of the system on delivery. This requirement would entail an extensive testing period, use of multiple test articles, and a likely focus on a single or small set of testing circumstances. To keep consumer risks low, the system might have been designed to have a greater reliability than required to reduce the chances of failing operational testing (although the procedure described in Chapter 7 on identifying a minimally acceptable level could reduce the need for this approach).

Unfortunately, in many cases, often because of fiscal constraints, producing an estimate of system reliability that has narrow confidence intervals may not be feasible, and thus it may be advisable to pursue additional opportunities for assessing operational reliability. Such an assessment could best be accomplished by having the last developmental test be operationally flavored to enhance prospects for combining information from operational and developmental tests. (This approach could also aid in the detailed planning for initial operational testing and evaluation.)

One way to determine in what way full-system developmental testing could be more operationally realistic relative to a reliability assessment would be to allow for early and persistent opportunities for users and operators to interact with the system. This approach would also support feedback to system improvement processes—which should be encouraged—whether within dedicated test events or undertaken more informally. Also, one or more focused follow-on tests could be conducted after the initial operational testing, allowing previously observed deficiencies and newly implemented redesigns or fixes to be examined. However, achieving a test of suitable length for reliability remains a challenge; thus, it is far better for the emphasis to be on developmental testing to produce a system whose reliability is already at the required level.[2]

---

[2]Some appeals to use sequential methods have been made to make testing more efficient. However, even though learning could be accommodated during operational testing as part of a sequential design, practical constraints (e.g., delaying scoring conferences until failure causes are fully understood) limit its use.

We argue above about the importance of modifying developmental test design to explore potentially problematic components or subsystems that have been found in contractor testing. The same argument applies to the relationship between developmental and operational testing. If a reliability deficiency is discovered in developmental testing and a system modification is made in response, then it may be useful to add some replications in operational testing to make sure that the modification was successful. We understand that operational test designs often attempt to mimic the profile of intended use represented in the operational mode summary/mission profile. However, such designs could often be usefully modified to provide information on problems raised in developmental testing, especially when such problems either have not been addressed by design modifications, or when such design modifications were implemented late in the process and so have not been comprehensively tested.

TEST DATA ANALYSIS

To greatly enhance the information available from the analysis of operational testing, it is important to collect and store reliability data on a per test article basis to enable the construction of complete histories (on operating times, operational mode summary/mission profile phase or event, failure times, etc.) for all individual articles being tested. If operational testing is done in distinct test events, then this ability to recreate histories should remain true across tests. Also, it would be valuable for data to be collected and stored to support analyses for distinct types of failure or failure modes, as mentioned in Chapter 7.

It is of critical importance that all failures are reported. If a failure is to be removed from the test data, then there needs to be an explanation that is fully documented and reviewable. It is also important that records are sufficiently detailed to support documented assessments of whether any observed failure should be discounted. Even if an event is scored as a "no test," it is important that observed failures are scrutinized for their potential to inform assessments concerning reliability growth.

In order to determine what reliability model is most appropriate to fit to the data from an operational test, it is often extremely useful to graphically display the timeline of failures (possibly disaggregated by failure type or by subsystem) to check whether the assumption of a specific type of life distribution model is appropriate (especially the exponential assumption). In particular, time trends that suggest the beginning of problems due to wear-out or various forms of degradation are important to identify. Crowder et al. (1991) has many useful ideas for graphical displays of reliability data. Such graphical displays are also extremely important to use in assisting decision makers in understanding what the results from developmental and operational testing indicate about the current level of performance. The U.S. Army Test Evaluation Command has some excellent graphical tools that are useful for this purpose (see, e.g., Cushing, 2012).

If the data have been collected and made accessible, then their comprehensive analysis can provide substantially more information to decision makers than a more cursory, aggregate analysis, so that decisions on promotion to full-rate production and other issues are made on the basis of a more complete understanding of the system's capabilities. In other words, the analysis of an operational test should not be focused solely on the comparison of the estimated key performance parameters to the system's requirements. Although such aggregate assessments are important because of their roles in decisions on whether the system has "passed" its operational test, there is other information that operational test data can provide that can be extremely important. For example, what if the new system was clearly superior to the current system in all but one of the proposed scenarios of use, but clearly inferior in that one? Such a determination could be important for deciding on tactics for the new system. Or what if the new system was superior to the existing system for all but one of the prototypes but distinctly inferior for that prototype? In this case, addressing manufacturing problems that occasionally resulted in poor quality items would be an important issue to explore. These are just two examples of a larger number of potential features of the performance of systems in such situations that can be discovered through detailed data analysis. In these two situations, there are statistical methods for quantifying heterogeneity across articles (e.g., random effects models) and describing performance across diverse conditions (e.g., regression models) that can be used to provide such assessments.

If there is a need to analytically combine developmental and operational test data to improve the precision of estimated reliability parameters, there are modeling challenges related to fundamental questions of comparable environmental conditions, operators and users, usage profiles, data collection and scoring rules, hardware/software configurations, interfaces, etc. (see discussion in Chapter 7). Such statistical challenges would confront any ex post facto attempts to formally combine operationally relevant results across these sorts of diverse testing events. (The specifics will vary across systems.) Although we strongly encourage attempts to address this challenge, such efforts should be considered difficult research problems for which only a few individual cases have been explored. Steffey et al. (2000) is one effort that we know of to address this problem, but their approach updates a single parameter, which we believe has limited applicability without proper extension. Also, again as mentioned above, the PREDICT technology can be considered an attempt to address this problem.

We also note that during an operational test event, a system's aging, fatigue, or degradation effects may be observable but may not have progressed to the point that an actual failure event is recorded. Examples include vehicle tread wear and machine gun barrel wear. Ignoring such information may prevent the diagnosis of the beginnings of wear-out or fatigue. When the effects are substantial, statistical modeling of degradation data can yield improved, operationally representative estimates of longer-term failure rates (see, e.g., Meeker and Escobar, 1998, Chs. 13 and 21; Escobar et al., 2003). The need for such analyses may be anticipated a priori or may arise from analysis of developmental test outcomes. Such analysis can be extremely important, because for some major performance-critical subsystems (e.g., aircraft engines) the development of empirical models for tracking reliability and projecting prudent replacement times can have profound cost and safety impacts. As results on operational reliability emerge from the initial operational test and evaluation (and possibly follow-on operational tests and evaluations), it is prudent to address these potential sources of information on lifetime ownership costs and factor these considerations into the planning for reliability growth opportunities. The flexible COHORT (COnsumption, HOlding, Repair, and Transportation) model of the Army Materiel Systems Analysis Agency (AMSAA) exemplifies this concept (see Dalton and Hall, 2010). Also, for repairable systems, displays of the failure distributions for new systems versus systems with varying numbers of repairs would be informative.

Finally, there are some estimates that are directly observed in operational testing, and there are some estimates that are model based and hence have a greater degree of uncertainty due to model misspecification. As the discussion above of life-cycle costs makes clear, whenever important inferences are clearly dependent on a model that is not fully validated, sensitivity analyses need to be carried out to provide some sense of the variability due to the assumed model form. Operational test estimates should either be presented to decision makers as being "demonstrated results" (i.e., directly observable in an operational test) or as "inferred estimates" for which the sensitivity to assumptions is assessed.

THE DT/OT GAP

It is well known that reliability estimates using data from developmental testing are often optimistic estimates of reliability under operationally relevant conditions. Figure 8-1 shows the ratio of developmental test and operational test reliability for Army acquisition category I systems from 1995-2004. (The 44 systems depicted were chosen from a database of Army systems created by the Army Test and Evaluation Command to satisfy the following criteria: (1) systems with both DT and OT results, (2) reporting the same reliability metric—some version of mean time between

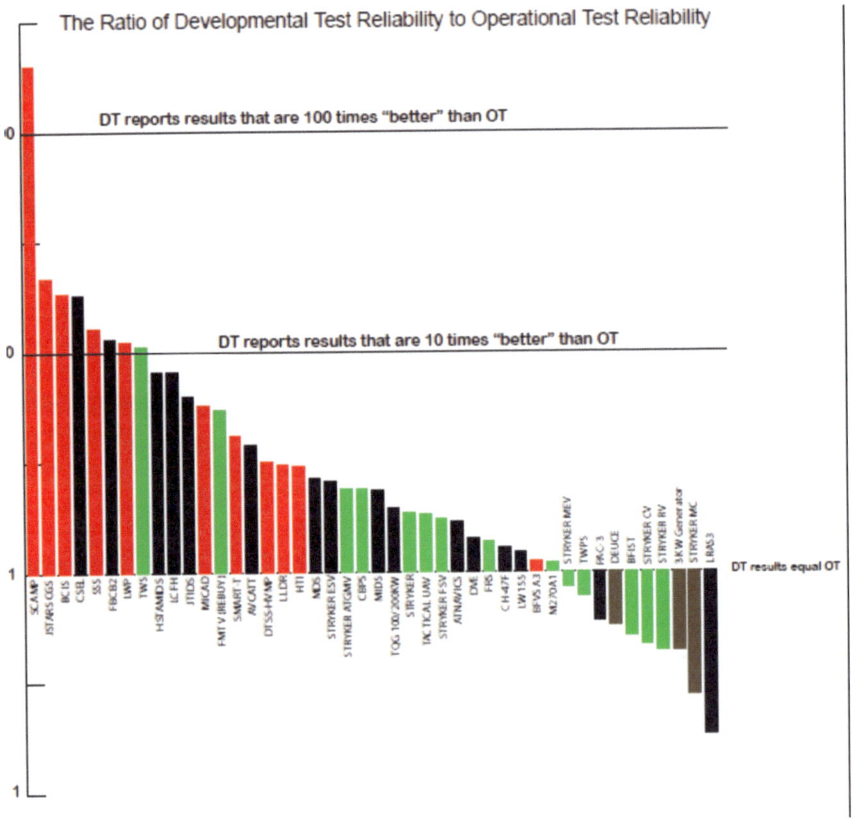

FIGURE 8-1 Comparison of developmental test and operational test reliability.
NOTES: DT = developmental testing, OT = operational testing. Red bars are systems where DT concluded reliability MET, but OT concluded NOT MET. Black bars are systems where DT and OT concluded NOT MET. Green bars are systems where DT and OT concluded MET. Brown bars are systems where DT concluded NOT MET, but OT concluded MET. The Y axis displays the log (reliability metric for DT/reliability metric for OT).
SOURCE: E. Seglie, Reliability in Developmental Test and Evaluation and Operational Test and Evaluation, p. 5). Unpublished.

failure, and (3) the DT and OT were relatively close in time.) This DT/OT gap is likely due to several aspects of developmental testing:

1. the use of highly trained users rather than users typical of operational use,
2. failure to include representation of enemy systems or countermeasures,

3. the scripted nature of the application of the system in which the sequence of events is often known to the system operators, and
4. representation of some system functions only through the use of modeling and simulation.

The last point above may be the reason that interface and interoperability issues often appear first in operational tests. Furthermore, as suggested by (3), the scenarios used in developmental testing often are unlike the scenarios used in operational testing, and it is therefore difficult to match the exercising of a system in developmental testing with the results of specific missions in operational testing (an issue we return to below).

The problem raised by the DT/OT gap is that systems that are viewed as having reliability close or equal to the required level are judged ready for operational testing when their operationally relevant reliability may be substantially smaller. As a result, either it will be necessary to find a large number of design defects during operational testing, a goal inconsistent with the structure of operational testing, or a system with deficient reliability will be promoted to full-rate production. (There is also, as stressed above, the inefficiency of making design changes so late in development.)

There are both design and analytic approaches to address the DT/OT gap. We start with design possibilities to address the four differences listed above (i.e., degree of training, representation of countermeasures and enemy forces, degree of scripting used, and representation of interfaces and interoperability). In terms of having users with typical training in tests, while we understand that it might be difficult to schedule such users for many developmental tests, we believe that systems, particularly software systems, can be beta-tested by asking units to try out systems and report back on reliability problems that are discovered. Second, for many types of enemy systems and countermeasures, various forms of modeling and simulation can provide some version of the stresses and strains that would result from taking evasive actions, etc. Third, relative to scripting, and closely related to the representation of enemy forces and countermeasures, there is often no reason why the system operator needs to know precisely the sequence of actions that will be needed. Finally, in terms of the representation of a system's functions that have not been completed, this is one reason for having a full-system, operationally relevant test prior to delivery to DoD for developmental testing.

In addition to design changes, the DT/OT gap could be analytically accounted for by estimating the size of the gap and adjusting developmental testing estimates accordingly. Input into the development of such models could be greatly assisted by the creation of the database we recommend (see Recommendation 24, in Chapter 10) that would support comparison of reliability estimates from developmental and operational testing.

At the panel's workshop, Paul Ellner of AMSAA commented on this challenge and what might be done to address it. He said that when the requirement is stated in terms of the mean time between failures, the developmental testing goal should be higher than the requirement because one needs to plan for a decrease in mean time between failures between developmental and operational tests because of both new failure modes surfaced by operational testing and higher failure rates for some failure modes that operational and developmental tests share.

Ellner said that a large DT/OT gap has three serious consequences:

1. It places the system at substantial risk of not passing the operational test.
2. It can lead to the fielding of systems that later require costly modifications to enhance mission reliability and reduce the logistics burden.
3. It is not cost-effective, nor may it be feasible, to attain sufficiently high reliability with respect to the developmental test environment to compensate for poor or marginal reliability with respect to potential operational failure modes.

After all, he said, one cannot attain better than 100 percent reliability in developmental testing. Furthermore, that could be meaningless with respect to reliability under operationally relevant use if developmental testing cannot elicit the same failure modes as those that appear in operational use.

For a real example, Ellner mentioned a developmental test for a vehicle with a turret. During the developmental test, the vehicle was stationary, the system was powered by base power, and contractor technicians operated the system. In operational testing, the vehicles were driven around, the system was powered by vehicle power, and military personnel operated the system. In the developmental test, 8 percent of the failure rate was due to problems with the turret. However, when the user profile was weighted to reflect the operational mode summary/mission profile, 23.9 percent of the failure rate was due to problems with the turret. Even more striking, in operational testing, 60.4 percent of the failure rate was due to the turret.

To reduce the size of this DT/OT reliability gap, Ellner proposed ways of either modifying the design of developmental test events, or modifying the test analysis. For the design of developmental testing, it could be based more closely on the operational use profile through the use of what Ellner called "balanced" testing, which is that the cumulative stress per time interval of test should closely match that of the operational mission profile.

In modifying the analysis, Ellner advocated a methodology that combines developmental and operational test data through use of Bayesian methods that leverage historical information for different types of systems with respect to their previously observed DT/OT gaps (see, e.g., Steffey et al., 2000). One difficulty with this approach is that not only are the scenarios of use changing from developmental testing to operational testing to the field, but also the systems themselves are changing as the design is refined as a result of test results. The only way to know whether such dynamics can be successfully dealt with is to make use of case studies and see how estimates from such models compare with reliability estimates based on field performance.

Ellner distinguished between different phases of developmental testing, because some functionalities are often not exercised in early developmental tests, but only in later stages of developmental testing. Also, later developmental testing often involves full-system tests, while early developmental testing often involves component-level and subsystem tests. Restricting attention to the comparison of late-stage developmental to operational testing, Ellner pointed out that there may be "normalizations" that can be used to reweight scenarios of use in developmental tests to match the scenarios faced in operational tests (or field use). That is, when feasible, one could try to reweight the developmental test scenarios to reflect what would have been the stress rate used in operational testing. For example, if only 10 percent of the test excursions were on rough terrain, but 80 percent of anticipated operational missions are over rough terrain, then the estimated occurrence rate of terrain-induced failures could be suitably weighted so that in effect 80 percent of the developmental testing time was in rough terrain.

The panel is strongly supportive of the use of these mitigation procedures for reducing the magnitude of the DT/OT gap. Furthermore, the panel recommends investigating the development of statistical models of the gaps (see Recommendation 24, in Chapter 10).

9

Software Reliability Growth

Somewhat analogous to the topics we have covered in previous chapters for hardware systems, this chapter covers software reliability growth modeling, software design for reliability, and software growth monitoring and testing.

Software reliability, like hardware reliability, is defined as the probability that the software system will work without failure under specified conditions and for a specified period of time (Musa, 1998). But software reliability differs in important respects from hardware reliability. Software reliability problems are deterministic in the sense that each time a specific set of inputs is applied to the software system, the result will be the same. This is clearly different from hardware systems, for which the precise moment of failure, and the precise cause of failure, can differ from replication to replication. In addition, software systems are not subject to wear-out, fatigue, or other forms of degradation.

In some situations, reliability errors are attributed to a full system and no distinction is made between subsystems or components, and this attribution is appropriate in many applications. However, as is the case for any failure mode, there are times when it is appropriate to use separate metrics and separate assessments of subsystem or component reliabilities (with respect to system structure as well as differentiating between software and hardware reliability), which can then be aggregated for a full-system assessment. This separate treatment is particularly relevant to software failures given the different nature of software and hardware reliability.

Chapter 4 on hardware reliability growth is primarily relevant to growth that occurs during full-system testing, which is relevant to the middle and later stages of developmental testing. In contrast, except for when the entire system is software, it is appropriate for software reliability growth to be primarily considered as a component-level concern, which would be addressed while the system is in development by the contractor, or at the latest, during the earliest stages of developmental testing. Therefore, the primary party responsible for software reliability is the contractor.

In this chapter we first discuss software reliability growth modeling as it has been generally understood and used in defense acquisition. We then turn to a new approach, metrics-based modeling: we describe the work that has been done and discuss how to build metrics-based prediction models. The last two sections of the chapter briefly consider testing and monitoring.

SOFTWARE RELIABILITY GROWTH MODELING

Classic Design Models

Software reliability growth models have, at best, limited use for making predictions as to the future reliability of a software system in development for several reasons. Most important, the pattern of reliability growth evident during the development of software systems is often not monotonic because corrections to address defects will at times introduce additional defects. Therefore, although the nonhomogeneous Poisson process model is one of the leading approaches to modeling the reliability of software (and hardware) systems in development, it often provides poor inferences and decision rules for the management of software systems in development.

Other deficiencies in such models relevant to software are the substantial dependence on time as a modeling factor, the dynamic behavior of software systems, the failure to take into consideration various environmental factors that affect software reliability when fielded, and hardware interactions. With respect to the dependence on time, it is difficult to create a time-based reliability model for software systems because it is highly likely that the same software system will have different reliability values in relation to different software operational use profiles. The dynamic behavior of software systems as a function of the environment of use, the missions employed, and the interactions with hardware components, all complicate modeling software reliability.

Siegel (2011, 2012) describes related complexities. Often metric-based models for software reliability, derived from a large body of recent research ranging from code churn, code complexity, code dependencies, testing coverage, bug information, usage telemetry, etc., have been shown to be effective predictors of code quality. Therefore, this discussion of software reliability growth models is followed by a discussion of the use of metric-based models, which we believe have important advantages as tools for predicting software reliability of a system.

A number of models of software reliability growth are available and represent a substantial proportion of the research on software reliability. They range from the simple Nelson model (Nelson, 1978), to more sophisticated hyper-geometric coverage-based models (e.g., Jacoby and Masuzawa, 1992), to component-based models and object-oriented models (e.g., Basili et al., 1996). Several reliability models use Markov chain techniques (e.g., Whittaker, 1992). Other models are based on the use of an operational profile, that is, a set of software operations and their probabilities of occurrence (e.g., Musa, 1998). These operational profiles are used to identify potentially critical operational areas in the software to signal a need to increase the testing effort in those areas. Finally, a large group of software reliability growth models are described by nonhomogenous Poisson processes (for a description, see Yamada and Osaki, 1985): this group includes Musa (e.g., Musa et al., 1987) and the Goel-Okumoto models (e.g., Goel and Okumoto, 1979).

Software reliability models can be classified broadly into seven categories (Xie, 1991):

1. Markov models: A model belongs to this class if its probabilistic assumption of the failure process is essentially a Markov process. In these models, each state of the software has a transition probability associated with it that governs the operational criteria of the software.
2. Nonhomogeneous Poisson process models: A model is in this class if the main assumption is that the failure process is described by a nonhomogeneous Poisson process. The main characteristic of this type of model is that there is a mean value function that is defined by the expected number of failures up to a given time.
3. Bayesian process models: In a Bayesian process model, some information about the software to be studied is available before the testing starts, such as the inherent fault density and defect information of previous releases. This information is then used in combination with the collected test data to more accurately estimate and make predictions about reliability.
4. Statistical data analysis methods: Various statistical models and methods are applied to software failure data. These models include time-series models, proportional hazards models, and regression models.
5. Input-domain based models: These models do not make any dynamic assumption about the failure processes. All possible input and output domains of the software are constructed and, on the

basis of the results of the testing, the faults in mapping between the input and output domains are identified. In other words, for a particular value in the input domain, either the corresponding value in the output domain is produced or a fault is identified.

6. Seeding and tagging models: These models are the same as capture-recapture methods based on data resulting from the artificial seeding of faults in a software system. The assessment of a test is a function of the percentage of seeded faults that remain undiscovered at the conclusion of the testing effort.
7. Software metrics models: Software reliability metrics, which are measures of the software complexity, are used in models to estimate the number of software faults remaining in the software.

A fair number of these classical reliability models use data on test failures to produce estimates of system (or subsystem) reliability. But for many software systems, developers strive for the systems to pass all the automated tests that are written, and there are often no measurable faults. Even if there are failures, these failures might not be an accurate reflection of the reliability of the software if the testing effort was not comprehensive. Instead, "no failure" estimation models, as described by Ehrenberger (1985) and Miller et al. (1992), may be more appropriate for use with such methodologies.

Another factor that affects classical software reliability models is that in software systems, the actual measurable product quality (e.g., failure rate) that is derived from the behavior of the system usually cannot be measured until too late in the life cycle to effect an affordable corrective action. When test failures occur in actual operation, the system has already been implemented. In general, a multiphase approach needs to be taken to collect the various metrics of the relevant subsystems at different stages of development, because different metrics will be estimable at different development phases and some can be used as early indicators of software quality (for a description of these approaches, see Vouk and Tai, 1993; Jones and Vouk, 1996). In Box 9-1, we provide short descriptions of the classical reliability growth models and some limitations of each approach.

Performance Metrics and Prediction Models

An alternative approach to reliability growth modeling for determining whether a software design is likely to lead to a reliable software system is to rely on performance metrics. Such metrics can be used in tracking software development and as input into decision rules on actions such as accepting subsystems or systems for delivery. In addition to such metrics, there has been recent work on prediction models, some of this stemming from the

BOX 9-1
Overview of Classical Software Reliability Growth Models

Nelson Model
This is a very simplistic model based on the number of test failures:

$$R = 1 - \frac{\hat{n}}{n},$$

where

- R is the system reliability
- $\hat{n}$ is the number of failures during testing
- n is the total number of testing runs.

A limitation of this model is that if no failures are available, the reliability becomes 100 percent, which might not always be the case. For details, see Nelson (1978).

Fault Seeding Models
In these models, faults are intentionally injected into the software by the developer. The testing effort is evaluated on the basis of how many of these injected defects are found during testing. Using the number of injected defects remaining, an estimate of the reliability based on the quality of the testing effort is computed using capture-recapture methods. A limitation of this model is that for most large systems, not all parts have the same reliability profile. The fault seeding could also be biased, causing problems in estimation. For details, see Schick and Wolvertone (1978) and Duran and Wiorkowski (1981).

Hypergeometric Distribution
This approach models overall system reliability by assuming that the number of faults experienced in each of several categories of test instance follows the hypergeometric distribution. However, if all the test cases pass, then there are no faults or failures to analyze. For details, see Tahoma et al. (1989).

Fault Spreading Model
In this model, the number of faults at each level (or testing cycle or stage) is used to make predictions about untested areas of the software. One limitation of the model is the need for data to be available early enough in the development cycle to affordably guide corrective action. For details, see Wohlin and Korner (1990).

Fault Complexity Model
This model ranks faults according to their complexity. The reliability of the system is estimated on the basis of the number of faults in each complexity level (high, moderate, low) of the software. For details, see Nakagawa and Hanata (1989).

continued

BOX 9-1 Continued

Littlewood-Verall Model
In this model, waiting times between failures are assumed to be exponentially distributed with a parameter assumed to have a prior gamma distribution. For details, see Littlewood and Verall (1971).

Jelinski-Moranda (JM) Model
In the JM model, the initial number of software faults is unknown but fixed, and the times between the discovery of failures are exponentially distributed. Based on this set-up, the JM model is modeled as a Markov process model. For details, see Jelinksi and Moranda (1972).

Bayesian Model for Fault Free Probability
This model deals with the reliability of fault-free software. Reliability at time t is assumed to have the following form:

$$R(t \mid \lambda, p) = (1 - p) + pe^{-\lambda t},$$

where λ is given by a prior gamma distribution and p (the probability that the software is not fault free) is given by a Beta distribution. Using these two parameters, a Bayesian model is constructed to estimate the reliability. For details, see Thompson and Chelson (1980).

Bayesian Model Using a Geometric Distribution
In this model, based on the number of test cases at the ith debugging instance for which a failure first occurs, the number of failures remaining at the current debugging instance is determined. For details, see Liu (1987).

Goel-Okumoto Model
This is a nonhomogeneous Poisson process model in which the mean of the distribution of the cumulative number of failures at time t is given by $m(t) = a(1 - e^{-bt})$, where a and b are parameters estimated from the collected failure data. For details, see Goel and Okumoto (1979).

S-Shaped Model
This is also a nonhomogeneous Poisson process model in which the mean of the distribution of the cumulative number of failures is given by $m(t) = a(1 -(1 + bt)e^{-bt}))$, where a is the expected number of faults detected, and b is the failure detection rate. For details, see Yamada and Osaki (1983).

Basic Execution Time Model
In this model, the failure rate function at time t is given by:

$$\lambda(t) = fK(N_0 - \mu(t)),$$

where

- f and K are parameters related to the testing phase
- N_0 is the assumed initial number of faults, and
- $\mu(t)$ is the number of faults corrected after t amount of testing.

A limitation of this model is that it cannot be applied when one does not have the initial number of faults and the failure rate function at execution time t. For details, see Musa (1975).

Logarithmic Poisson Model
This model is related to the basic execution time model (above). However, here the failure rate function is given by:

$$\lambda(t) = \lambda_0 e^{-\phi\mu(t)}$$

where λ_0 is the initial failure intensity, and ϕ is the failure intensity decay parameter. For details, see Musa and Okumoto (1984).

Duane Model (Weibull Process Model)
This is a nonhomogeneous Poisson process model with mean function

$$m(t) = \left(\frac{t}{\alpha}\right)^{\beta}.$$

The two parameters, α and β, are estimated using failure time data. For details, see Duane (1964).

Markov Models
Markov models require transition probabilities from state to state where the states are defined by the current values of key variables that define the functioning of the software system. Using these transition probabilities, a stochastic model is created and analyzed for stability. A primary limitation is that there can be a very large number of states in a large software program. For details, see Whittaker (1992).

Fourier Series Model
In this model, fault clustering is estimated using time-series analysis. For details, see Crow and Singpurwalla (1984).

Input Domain-Based Models
In these models, if there is a fault in the mapping of the space of inputs to the space of intended outputs, then that mapping is identified as a potential fault to be rectified. These models are often infeasible because of the very large number of possibilities in a large software system. For details, see Bastani and Ramamoorthy (1986) and Weiss and Weyuker (1988).

work of McCabe (1976) and more recent work along similar lines (e.g., Ostrand et al., 2005; Weyuker et al., 2008).

It is quite likely that for broad categories of software systems, there already exist prediction models that could be used earlier in development than performance metrics for use in tracking and assessment. It is possible that such models could also be used to help identify better performing contractors at the proposal stage. Further, there has been a substantial amount of research in the software engineering community on building generalizable prediction models (i.e., models trained in one system to be applied to another system); an example of this approach is given in Nagappan et al. (2006). Given the benefits from earlier identification of problematic software, we strongly encourage the U.S. Department of Defense (DoD) to stay current with the state of the art in software reliability as is practiced in the commercial software industry, with increased emphasis on data analytics and analysis. When it is clear that there are prediction models that are broadly applicable, DoD should consider mandating their use by contractors in software development.

A number of metrics have been found to be related to software system reliability and therefore are candidates for monitoring to assess progress toward meeting reliability requirements. These include code churn, code complexity, and code dependencies (see below).

We note that the course on reliability and maintainability offered by the Defense Acquisition University lists 10 factors for increasing software reliability and maintainability:

1. good statement of requirements,
2. use of modular design,
3. use of higher-order languages,
4. use of reusable software,
5. use of a single language,
6. use of fault tolerance techniques,
7. use of failure mode and effects analysis,
8. review and verification through the work of an independent team,
9. functional test-debugging of the software, and
10. good documentation.

These factors are all straightforward to measure, and they can be supplied by the contractor throughout development.

METRICS-BASED MODELS

Metrics-based models are a special type of software reliability growth model that have not been widely used in defense acquisition. These are software reliability growth models that are based on assessments of the change in software metrics that are considered to be strongly related to system reliability. The purpose of this section is to provide an understanding of when metrics-based models are applicable during software development. The standard from the International Organization for Standards and the International Electrotechnical Commission 1498-1 states that "internal metrics are of little value unless there is evidence that they are related to some externally visible quality." Internal metrics have been shown to be useful, however, as early indicators of externally visible product quality when they are related (in a statistically significant and stable way) to the field quality (reliability) of the product (see Basili et al., 1996).

The validation of such internal metrics requires a convincing demonstration that the metric measures what it purports to measure and that the metric is associated with an important external metric, such as field reliability, maintainability, or fault-proneness (for details, see El-Emam, 2000). Software fault-proneness is defined as the probability of the presence of faults in the software. Failure-proneness is the probability that a particular software element will fail in operation. The higher the failure-proneness of the software, logically, the lower the reliability and the quality of the software produced, and vice versa.

Using operational profiling information, it is possible to relate generic failure-proneness and fault-proneness of a product. Research on fault-proneness has focused on two areas: the definition of metrics to capture software complexity and testing thoroughness and the identification of and experimentation with models that relate software metrics to fault-proneness (see, e.g., Denaro et al., 2002). While software fault-proneness can be measured before deployment (such as the count of faults per structural unit, e.g., lines of code), failure-proneness cannot be directly measured on software before deployment.

Five types of metrics have been used to study software quality: (1) code churn measures, (2) code complexity measures, (3) code dependencies, (4) defect or bug data, and (5) people or organizational measures. The rest of this section, although not comprehensive, discusses the type of statistical models that can be built using these measures.

Code Churn

Code churn measures the changes made to a component, file, or system over some period of time. The most commonly used code churn measures are the number of lines of code that are added, modified, or deleted. Other churn measures include *temporal* churn (churn relative to the time of release of the system) and *repetitive* churn (frequency of changes to the same file or component). Several research studies have used code churn as an indicator of code quality (fault- or failure-proneness and, by extension, reliability). Graves et al. (2000) predicted fault incidences using software change history on the basis of a time-damping model that used the sum of contributions from all changes to a module, in which large or recent changes contributed the most to fault potential. Munson and Elbaum (1998) observed that as a system is developed, the relative complexity of each program module that has been altered will change. They studied a software component with 300,000 lines of code embedded in a real-time system with 3,700 modules programmed in C. Code churn metrics were found to be among the most highly correlated with problem reports.

Another kind of code churn is debug churn, which Khoshgoftaar et al. (1996) define as the number of lines of code added or changed for bug fixes. The researchers' objective was to identify modules in which the debug code churn exceeded a threshold in order to classify the modules as fault-prone. They studied two consecutive releases of a large legacy system for telecommunications that contained more than 38,000 procedures in 171 modules. Discriminant analysis identified fault-prone modules on the basis of 16 static software product metrics. Their model, when used on the second release, showed type I and type II misclassification rates of 21.7 percent and 19.1 percent, respectively, and an overall misclassification rate of 21.0 percent.

Using information on files with status new, changed, and unchanged, along with other explanatory variables (such as lines of code, age, prior faults) as predictors in a negative binomial regression equation, Ostrand et al. (2004) successfully predicted the number of faults in a multiple release software system. Their model had high accuracy for faults found in both early and later stages of development.

In a study on Windows Server 2003, Nagappan and Ball (2005) demonstrated the use of relative code churn measures (normalized values of the various measures obtained during the evolution of the system) to predict defect density at statistically significant levels. Zimmermann et al. (2005) mined source code repositories of eight large-scale open source systems (IBM Eclipse, Postgres, KOffice, gcc, Gimp, JBoss, JEdit, and Python) to predict where future changes would take place in these systems. The top three recommendations made by their system identified a correct location for future change with an accuracy of 70 percent.

Code Complexity

Code complexity measures range from the classical cyclomatic complexity measures (see McCabe, 1976) to the more recent object-oriented metrics, one of which is known as the CK metric suite after its authors (see Chidamber and Kemerer, 1994). McCabe designed cyclomatic complexity as a measure of the program's testability and understandability. Cyclomatic complexity is adapted from the classical graph theoretical cyclomatic number and can be defined as the number of linearly independent paths through a program. The CK metric suite identifies six object-oriented metrics:

1. weighted methods per class, which is the weighted sum of all the methods defined in a class;
2. coupling between objects, which is the number of other classes with which a class is coupled;
3. depth of inheritance tree, which is the length of the longest inheritance path in a given class;
4. number of children, which is the count of the number of children (classes) that each class has;
5. response for a class, which is the count of the number of methods that are invoked in response to the initiation of an object of a particular class; and
6. lack of cohesion of methods, which is a count of the number of method pairs whose similarity is zero minus the count of method pairs whose similarity is not zero.

The CK metrics have also been investigated in the context of fault-proneness. Basili et al. (1996) studied the fault-proneness in software programs using eight student projects. They found the first five object-oriented metrics listed above were correlated with defects while the last metric was not. Briand et al. (1999) obtained somewhat related results. Subramanyam and Krishnan (2003) present a survey on eight more empirical studies, all showing that object-oriented metrics are significantly associated with defects. Gyimóthy et al. (2005) analyzed the CK metrics for the Mozilla codebase and found coupling between objects to be the best measure in predicting the fault-proneness of classes, while the number of children was not effective for fault-proneness prediction.

Code Dependencies

Early work by Pogdurski and Clarke (1990) presented a formal model of program dependencies based on the relationship between two pieces of code inferred from the program text. Schröter et al. (2006) showed that such dependencies can predict defects. They proposed an alternate way of predicting failures for Java classes. Rather than looking at the complexity of a class, they looked exclusively at the components that a class uses. For Eclipse, the open source integrated development environment, they found that using compiler packages resulted in a significantly higher failure-proneness (71 percent) than using graphical user interface packages

(14 percent). Zimmermann and Nagappan (2008) built a systemwide code dependency graph of Windows Server 2003 and found that models built from (social) network measures had accuracy of greater than 10 percentage points in comparison with models built from complexity metrics.

Defect Information

Defect growth curves (i.e., the rate at which defects are opened) can also be used as early indicators of software quality. Chillarege et al. (1991) at IBM showed that defect types could be used to understand net reliability growth in the system. And Biyani and Santhanam (1998) showed that for four industrial systems at IBM there was a very strong relationship between development defects per module and field defects per module. This approach allows the building of prediction models based on development defects to identify field defects.

People and Social Network Measures

Meneely et al. (2008) built a social network between developers using churn information for a system with 3 million lines of code at Nortel Networks. They found that the models built using such social measures revealed 58 percent of the failures in 20 percent of the files in the system. Studies performed by Nagappan et al. (2008) using Microsoft's organizational structure found that organizational metrics were the best predictors for failures in Windows.

BUILDING METRICS-BASED PREDICTION MODELS

In predicting software reliability with software metrics, a number of approaches have been proposed. Logistic regression is a popular technique that has been used for building metric-based reliability models. The general form of a logistic regression equation is given as follows:

$$\Pr(\pi) = \frac{e^{c+a_1X_1+a_2X_2+\ldots}}{1+e^{c+a_1X_1+a_2X_2+\ldots}},$$

where c, a_1, and a_2 are the logistic regression parameters and X_1, X_2, . . . are the independent variables used for building the logistic regression model. In the case of metrics-based reliability models, the independent variables can be any of the (combination of) measures ranging from code churn and code complexity to people and social network measures.

Another common technique used in metrics-based prediction models is a support vector machine (for details, see Han and Kamber, 2006). For a quick overview of this technique, consider a two-dimensional training set with two classes as shown in Figure 9-1. In part (a) of the figure, points representing software modules are either defect-free (circles) or have defects (boxes). A support vector machine separates the data cloud into two sets by searching for a maximum marginal hyperplane; in the two-dimensional case, this hyperplane is simply a line. There are an infinite number of possible hyperplanes in part (a) of the figure that separate the two groups. Support vector machines choose the hyperplane with the margin that gives the largest separation between classes. Part (a) of the figure shows a hyperplane with a small margin; part (b) shows one with the maximum margin. The maximum margin is defined by points from the training data—these "essential" points are also called support vectors; in part (b) of the figure they are indicated in bold.

Support vector machines thus compute a decision boundary, which is used to classify or predict new points. One example is the triangle in part (c) of Figure 9-1. The boundary shows on which side of the hyperplane the new software module is located. In the example, the triangle is below the hyperplane; thus it is classified as defect free.

Separating data with a single hyperplane is not always possible. Part (d) of Figure 9-1 shows an example of nonlinear data for which it is not possible to separate the two-dimensional data with a line. In this case, support vector machines transform the input data into a higher dimensional space using a nonlinear mapping. In this new space, the data are then linearly separated (for details, see Han and Kamber, 2006). Support vector machines are less prone to overfitting than some other approaches because the complexity is characterized by the number of support vectors and not by the dimensionality of the input.

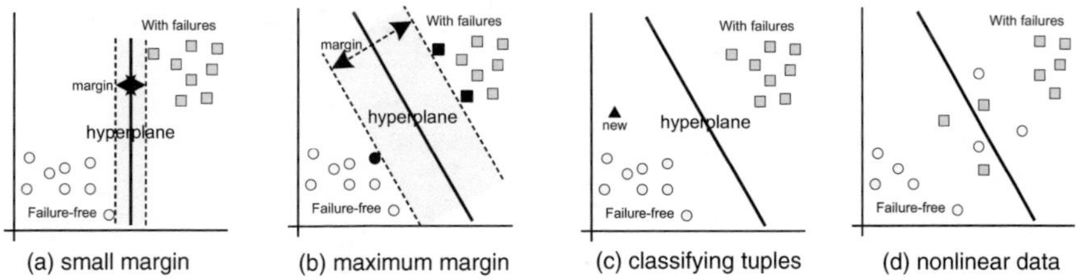

FIGURE 9-1 Support vector machines: overview.
NOTE: See text for discussion.

Other techniques that have been used instead of logistic regression and support vector machines are discriminant analysis and decision and classification trees.

Drawing general conclusions from empirical studies in software engineering is difficult because any process is highly dependent on a potentially large number of relevant contextual variables. Consequently, the panel does not assume a priori that the results of any study will generalize beyond the specific environment in which it was conducted, although researchers understandably become more confident in a theory when similar findings emerge in different contexts.

Given that software is a vitally important aspect of reliability and that predicting software reliability early in development is a severe challenge, we suggest that DoD make a substantial effort to stay current with efforts employed in industry to produce useful predictions.

TESTING

There is a generally accepted view that it is appropriate to combine software failures with hardware failures to assess system performance in a given test. However, in this section we are focusing on earlier non-system-level testing in developmental testing, akin to component-level testing for hardware. The concern is that if insufficient software testing is carried out during the early stages of developmental testing, then addressing software problems discovered in later stages of developmental testing or in operational testing will be much more expensive.[1]

As discussed in National Research Council (2006), to adequately test software, given the combinatorial complexity of the sequence of statements activated as a function of possible inputs, one is obligated to use some form of automated test generation, with high code coverage assessed using one of the various coverage metrics proposed in the research literature. This is necessary both to discover software defects and to evaluate the reliability of the software component or subsystem. However, given the current lack of software engineering expertise accessible in government developmental testing, the testing that can be usefully carried out, in addition to the testing done for the full system, is limited. Consequently, we recommend that the primary testing of software components and subsystems be carried out by the developers and carefully documented and reported to DoD and that contractors provide software that can be used to run automated tests of the component or subsystem (Recommendation 14, in Chapter 10).

[1]By software system, we mean any system that is exclusively software. This includes information technology systems and major automated information systems.

If DoD acquires the ability to carry out automated testing, then there are model-based techniques, including those developed by Poore (see, e.g., Whittaker and Poore, 1993), based on profiles of user inputs, that can provide useful summary statistics about the reliability of software and its readiness for operational test (for details, see National Research Council, 2006).

Finally, if contractor code is also shared with DoD, then DoD could validate some contractor results through the use of fault injection (seeding) techniques (see Box 9-1, above). However, operational testing of a software system can raise an issue known as fault masking, whereby the occurrence of a fault prevents the software system from continuing and therefore misses faults that are conditional on the previous code functioning properly. Therefore, fault seeding can fail to provide unbiased estimates in such cases. The use of fault seeding could also be biased in other ways, causing problems in estimation, but there are various generalizations and extensions of the technique that can address these various problems. They include explicit recognition of order constraints and fault masking, Bayesian constructs that provide profiles for each subroutine, and segmenting system runs.

MONITORING

One of the most important principles found in commercial best practices is the benefit from the display of collected data in terms of trend charts to track progress. Along these lines, Selby (2009) demonstrates the use of analytics dashboards in large-scale software systems. Analytics dashboards provide easily interpretable information that can help many users, including front-line software developers, software managers, and project managers. These dashboards can cater to a variety of requirements: see Figure 9-2. Several of the metrics shown in the figure, for example, the trend of post-delivery defects, can help assess the overall stability of the system.

Selby (2009) states that organizations should define data trends that are reflective of success in meeting software requirements so that, over time, one could develop statistical tests that could effectively discriminate between successful and unsuccessful development programs. Analytics dashboards can also give context-specific help, and the ability to drill down to provide further details is also useful: see Figure 9-3 for an example.

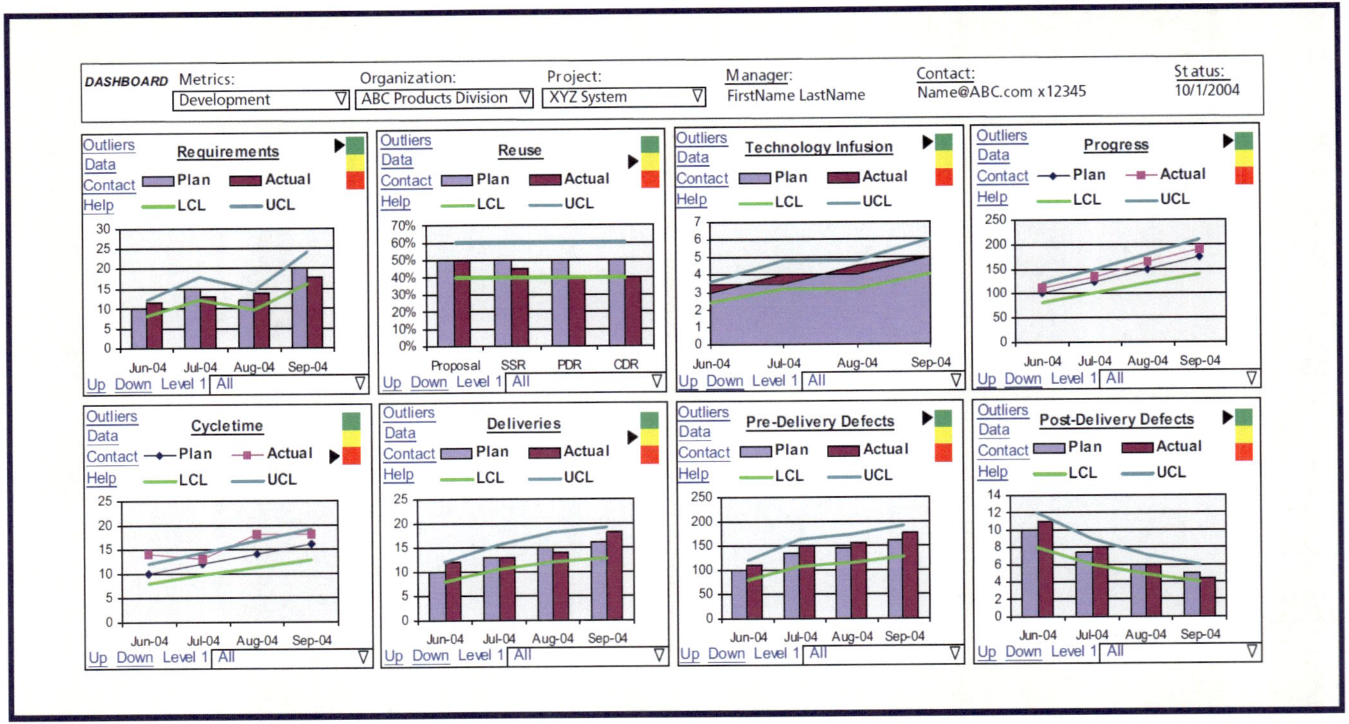

FIGURE 9-2 Analytics dashboards.
SOURCE: Selby (2009, p. 42). Reprinted with permission.

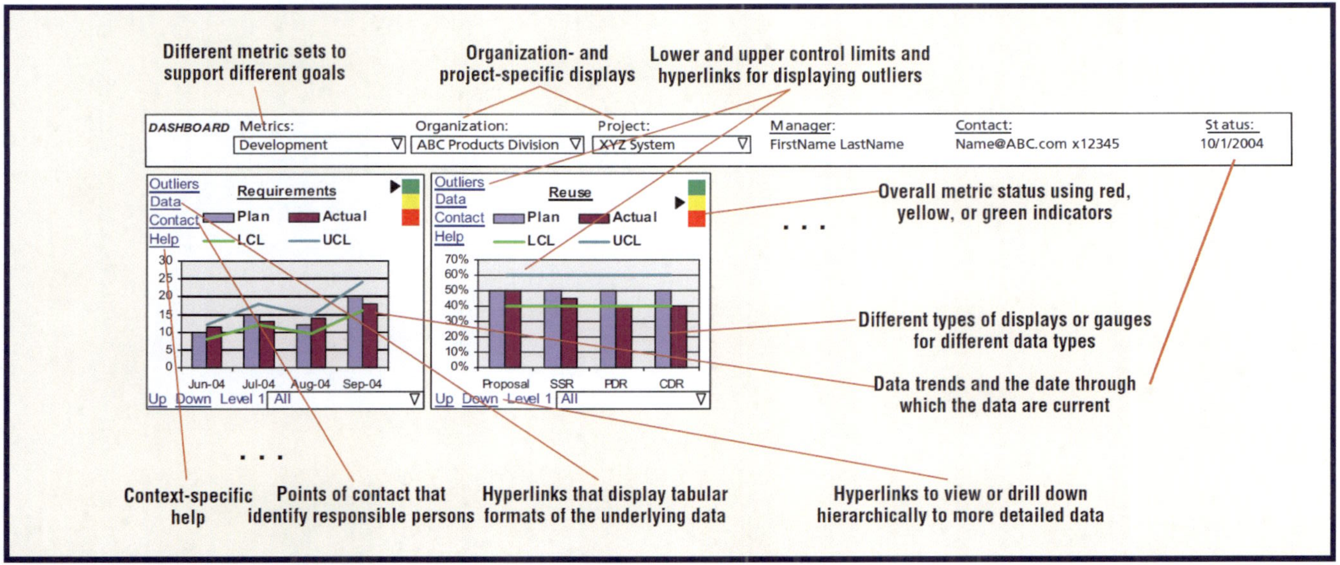

FIGURE 9-3 Example of a specific context of an analytics dashboard.
SOURCE: Selby (2009). Reprinted with permission.

10

Conclusions and Recommendations

Producing a reliable defense system is predicated on using proper engineering techniques throughout system development, beginning before program initiation, through delivery of prototypes, to fielding of the system. To start, one must develop requirements that are both technically achievable and measurable and testable. In addition, they need to be cost-effective when considering life-cycle costs.

Once reasonable requirements have been determined, the development of reliable defense systems depends on having an adequate budget and time to build reliability in at the design stage and then to refine the design through testing that is focused on reliability. We make several recommendations geared toward ensuring the allocation of sufficient design and test resources in support of the development of reliable defense systems. We also offer recommendations on information sharing and other factors related to the oversight of the work of contractors and subcontractors, the acceptance of prototypes from contractors, developmental testing, reliability growth modeling, and the collection and analysis of data throughout system development.

The panel's analysis and recommendations to the U.S. Department of Defense (DoD) cover the many steps and aspects of the acquisition process, presented in this chapter in roughly chronological order: analysis of alternatives; request for proposals; an outline reliability demonstration plan; raising the priority of reliability; design for reliability; reliability growth testing; design changes; information on operational environments; acquisition contracts; delivery of prototypes for developmental testing; developmental testing; and intermediate reliability goals.

We note that in several of our recommendations the panel designates the Under Secretary of Defense for Acquisition, Technology, and Logistics (USD AT&L) as the implementing agent. This designation reflects the panel's understanding of DoD's acquisition process and regulations and the flow of authority from USD AT&L, first through the Assistant Secretary of Defense for Networks and Information Integration (ASD NII) and the Director of Operational Test and Evaluation (DOT&E) and then through the component acquisition authorities (of each service) and program executive officers to program managers.[1]

We also note that some of our recommendations are partly redundant with existing acquisition procedures and regulations: our goal in including them is to emphasize their importance and to encourage more conscientious implementation.

ANALYSIS OF ALTERNATIVES

The defense acquisition process begins when DoD identifies an existing military need that requires a materiel solution. The result can be a request either for the development of a new defense system or the modification of an existing one. Different suggestions for addressing the need are compared in an "analysis of alternatives." This document contains the missions that a proposed system is intended to carry out and the conditions under which the proposed system would operate. Currently, the analysis of alternatives does not necessarily include the possible effects of system reliability on life-cycle costs (although many such analyses do). Clearly, those costs do need to be considered in the decision on whether to proceed.

After there is a decision to proceed, reliability requirements are first introduced and justified in the RAM-C (reliability, availability, maintainability, and cost) document, which lays out the reliability requirements for the intended system and contains the beginnings of a reliability model justifying that the reliability requirement is technically feasible.

If it is decided to develop a new defense system, possible contractors from industry are solicited using a request for proposal (RFP), which is based on both the analysis of alternatives and the RAM-C document. RFPs describe the system's capabilities so that potential bidders fully understand what is requested. RFPs specify the intended missions the system needs to successfully undertake, the conditions under which the system will operate

[1]DoD 5000.02 states, Program Managers for all programs shall formulate a viable Reliability, Availability and Maintainability strategy that includes a reliability growth program. Our recommendations, if implemented, will expand on this existing requirement and effect the work and authority of program managers and test authorities, but regulatory change is the responsibility of USD (AT&L), together with ASD (NII) and DOT&E.

and be maintained during its lifetime, the requirements that the system needs to satisfy, and what constitutes a system failure. An RFP also contains, from the RAM-C document, the beginning of a reliability model so that the contractor can understand how DoD can assert that the reliability requirement is achievable.

RFPs generate proposals, and these need to be evaluated, among other criteria, to assess whether the contractor is likely to produce a reliable system. Therefore, proposals need to be explicit about the design tools and testing, including a proposed testing schedule that will be funded in support of the production of a reliable system. When DoD selects a winning proposal, an acquisition contract is negotiated. This contract is critical to the entire process. Such contracts provide for the level of effort devoted to reliability growth, and the degree of interaction between the contractor and DoD, including the sharing of test and other information that inform DoD as to what the system in development is and is not capable of doing.

In making our recommendations, we consider first the analysis of alternatives. As noted above, there is currently no obligation in the analysis of alternatives to consider the impact of the reliability of the proposed system on mission success and life-cycle costs. Because such considerations could affect the decision as to whether to go forward with a new acquisition program, they should be required in every analysis of alternatives.

RECOMMENDATION 1 The Under Secretary of Defense for Acquisition, Technology, and Logistics should ensure that all analyses of alternatives include an assessment of the relationships between system reliability and mission success and between system reliability and life-cycle costs.

The next stage in the acquisition process is the setting of reliability requirements. Although these requirements should not necessarily be shared at the RFP stage, they are needed internally—even prior to the issuance of an RFP—to begin the process of justification and assessment of feasibility. The RAM-C report should justify the reliability requirements by showing that they are either necessary in order to have a high probability of successfully carrying out the intended missions or by showing that they are necessary to limit life-cycle costs.

In addition, the RAM-C report should include an estimate of the acquisition costs and an assessment of their uncertainty, which should include as a component the estimated life-cycle costs and an assessment of their uncertainty, with life-cycle costs expressed as a function of system reliability. (It is understood that life-cycle costs are a function of many other system characteristics than its reliability.) In addition, the RAM-C report should provide support for the assertion that the reliability requirements are technically feasible, measurable, and testable. (A requirement is measurable if there is a metric that underlies the requirement that is objectively determined, and it is testable if there is a test that can objectively discriminate between systems that have and have not achieved the requirement.)

DoDI 5000.02 requires

> [a] preliminary Reliability, Availability, Maintainability and Cost Rationale (RAM-C) Report in support of the Milestone A decision. This report provides a quantitative basis for reliability requirements, and improves cost estimates and program planning. This report will be attached to the SEP at Milestone A, and updated in support of the Development RFP Release Decision Point, Milestone B, and Milestone C. . . . [The RAM-C report] documents the rationale behind the development of the sustainment requirements along with underlying assumptions. Understanding these assumptions and their drivers will help warfighters, combat developers, and program managers understand the basis for decisions made early in a program. When the requirements and underlying assumptions are not clearly documented, the project may be doomed to suffer from subsequent decisions based on incorrect assumptions.

We are aware of reliability requirements for proposed new defense systems that have been technically infeasible or that have not reflected a cost-efficient approach to acquisition. Furthermore, reliability requirements have at times not been measurable or testable. To address these deficiencies, DoD should be obligated to include technical justifications in the RAM-C document that support these assertions in a manner that most experts in the field would find persuasive. Given that estimates of life-cycle costs require considerable technical expertise to develop, it is important to ensure that such assessments are made by appropriate experts in reliability engineering. Furthermore, the assessment as to whether requirements are achievable, measurable, and testable also requires considerable expertise with respect to the proposed system. To ensure that the required report about the reliability requirements reflects input from people with the necessary expertise, DoD should require that an external panel examines the arguments behind such assertions prior to the issuance of an RFP. That assessment of reliability requirements should be delivered to the Joint Requirements Oversight Council (JROC) or, as appropriate, its Component analog. This assessment should also contain an evaluation of the feasibility of acquiring the system within the specified cost and time schedule. The JROC, based on this technical report and the external assessment of it, should be the deciding authority on whether or not DoD proceeds to issue an RFP for the system.[2]

[2]In forming these expert committees, it is important that the relevant requirements officer is either a member or is asked to present any relevant work on the development of the reliability requirement.

RECOMMENDATION 2 Prior to issuing a request for proposals, the Office of the Under Secretary of Defense for Acquisition, Technology, and Logistics should issue a technical report on the reliability requirements and their associated justification. This report should include the estimated relationship between system reliability and total acquisition and life-cycle costs and the technical justification that the reliability requirements for the proposed new system are feasible, measurable, and testable. Prior to being issued, this document should be reviewed by a panel with expertise in reliability engineering, with members from the user community, from the testing community, and from outside of the service assigned to the acquisition. We recognize that before any development has taken place these assessments are somewhat guesswork and it is the expectation that as more about the system is determined, the assessments can be improved. Reliability engineers of the services involved in each particular acquisition should have full access to the technical report and should be consulted prior to the finalization of the RFP.

REQUESTS FOR PROPOSALS

As argued above, requests for proposals should include reliability requirements and their justification—highlighting the reliability goals for specific subsystems that are believed to be keys to system reliability—by demonstrating that they are necessary either to have a high probability of successfully carrying out the intended missions or by showing that a reliability level is necessary to limit life-cycle costs.[3] We acknowledge here, too, that prior to any system development, the assessment of feasibility and the linking of the level of system reliability to life-cycle costs is at best informed guesswork. But, absent assessments of feasibility, requirements could be optimistic dreams. And absent linking the requirements to reliability-driven life-cycle costs, the decision could be made to, say, modestly reduce the cost of the system through what would be perceived as a modest reduction in system reliability, and as a result producing a system that is substantially more expensive to field due to the increased life-cycle costs.

On the basis of the RAM-C document, the RFP should include a rough estimate of the acquisition costs and an assessment of their uncertainty, which should include as a component the estimated life-cycle costs and an assessment of their uncertainty, with life-cycle costs expressed as a function of system reliability. The RFP needs to provide support for the assertion

[3]Sometimes, system requirements are initially expressed optimistically to generate early support for the system. This is clearly counterproductive for many reasons, and the panel's recommendation to provide technical justification in the RFP may help to eliminate this practice.

that the reliability requirements are technically feasible by reporting estimated levels for specific subsystems thought to contribute substantially to system reliability, that have either appeared in fielded systems or for which estimates are otherwise available, and the assertion that the reliability requirements are measurable and testable.

Clearly, analyses of feasibility and estimates of life-cycle costs as a function of system reliability are likely to be revised as development proceeds by the contractor. But including initial analyses and assessments of these quantities in the RFP will help to demonstrate the high priority given to such considerations. As the system design matures, such analyses and assessments will improve. As a start to this improvement, part of the proposal that is produced in response to the RFP from the contractor should be the contractor's review of the government's initial reliability assertions and the degree to which they are in accordance or they differ (with rationale)—and the consequence of such on the contractor's proposal for designing, building, and testing the system.

In situations in which new technology is involved, DoD may instead issue a request for information to engage knowledgeable people from industry in the process of preparing the report on requirements. If new or developing technology may be needed in the system, the process of evolutionary acquisition needs to be considered.[4] In this case, necessary, achievable, measurable, and testable reliability requirements for the system during each acquisition spiral need to be specified and justified.

Even when assessments of technical feasibility are made with due diligence, it may be that during development the reliability requirements turn out to be technically infeasible. This possibility can become clear as a result of the collection of new information about the reliability of components and other aspects of the system design through early testing. Similarly, an argument about whether to increase system reliability beyond what is necessary for mission success in order to reduce life-cycle costs could need reconsideration as a result of the refinement of estimates of the costs of repair and replacement parts.

If the requirement for reliability turns out not to be technically feasible, it could have broad implications for the intended missions, life-cycle costs, and other aspects of the system. Therefore, when a request is made by the contractor for a modification of reliability requirements, there is again a need for a careful review and issuance of an abbreviated version of the analysis of alternatives and the above report on reliability requirements, with input from the appropriate experts. In addition to updating the

[4]For a description of this process, see DoD Instruction 5000.02, Operation of the Defense Acquisition System, available at http://www.dtic.mil/whs/directives/corres/pdf/500002_interim.pdf [December 2013].

analysis of alternatives, if necessary, the RAM-C and associated logistics documents would need to be updated to identify and show the impacts of the reliability changes.

RECOMMENDATION 3 Any proposed changes to reliability requirements by a program should be approved at levels no lower than that of the service component acquisition authority. Such approval should consider the impact of any reliability changes on the probability of successful mission completion as well as on life-cycle costs.

It is not uncommon for the DoD requirements generation process to establish one or more reliability requirements that differ from the reliability requirements agreed to in the acquisition contract. This can be due to the difference between mean time between failures in a laboratory setting and mean time between failures in an operational setting, or it can be due to negotiations between DoD and the contractor. In the first instance, these differences are due to the specifics of the testing strategy. To address this, we suggest that DoD archive the history of the development of the initial reliability requirement in the RFP and how that initial requirement evolved throughout development and even in initial fielding and subsequent use.

AN OUTLINE RELIABILITY DEMONSTRATION PLAN

Knowing the design of the tests that will be used to evaluate a system in development is an enormous help to developers in understanding the missions and the stresses the system is expected to face. Given the importance of conveying such information as early as possible to developers, RFPs should provide an early overview of what will later be provided in much greater detail in an outline reliability demonstration or development plan. With respect to reliability, a test and evaluation master plan (TEMP) provides the types and numbers of prototypes, hours, and other characteristics of various test events and schedules that will take place during government testing. And with respect to reliability assessment, a TEMP provides information on any acceleration used in tests, the associated evaluations resulting from tests, and, overall, how system reliability will be tracked in a statistically defendable way. So a TEMP provides a description of the various developmental and operational tests that will be used to identify flaws in system design and those tests that will be used to evaluate system performance. A TEMP also describes system failure and specifies how reliability is scored at test events and at design reviews.

It would be premature to lay out a TEMP in the RFP for a proposed new defense acquisition. However, having some idea as to the testing that is expected to be done to support reliability growth and to assess reliability performance would be extremely useful in making decisions on system design. We therefore call on DoD to produce a new document, which we call an outline reliability demonstration plan, to be included in the RFP and serve as the overview of the TEMP for reliability, providing as much information as is available concerning how DoD plans to evaluate system performance—for present purposes, the evaluation of reliability growth. The outline should specify the extent of the tests (e.g., total hours, number of replications), the test conditions, and the metrics used. The outline should also include the pattern of reliability growth anticipated at various stages of development.[5]

Preliminary reliability levels that can serve as intermediate targets would be available early on since there is some empirical evidence as to the degree of reliability growth that can be expected to result from a test-analyze-fix-test period of a certain length for various kinds of systems (see Chapter 6). An outline reliability demonstration plan should also indicate how such comparisons will be used as input to decisions on the promotion of systems to subsequent stages of development—a specified threshold needs to include a buffer to reflect the sample size of such tests in order to keep the producer's risk low.

As with the technical report on reliability requirements recommended above and because an outline reliability demonstration plan also has substantial technical content, it should be reviewed by an expert panel prior to its inclusion in an RFP. This expert panel should include reliability engineers and system users, members from the testing community, and members from outside of the service responsible for system acquisition. This expert panel should deliver a report reviewing the adequacy of the outline reliability demonstration plan, and it should include an assessment as to whether or not the system can likely be acquired within the specified cost and time schedule. Based on the technical report on reliability requirements and the outline reliability demonstration plan, the JROC would decide whether or not DoD will proceed to issue an RFP for the system.

RFPs currently contain a systems engineering plan, which lays out the methods by which all system requirements having technical content, technical staffing, and technical management are to be implemented on a program, addressing the government and all contractor technical efforts. Therefore the systems engineering plan is a natural location for this additional material on reliability test and evaluation that we argue for inclusion in RFPs.

---

[5]DoD may also wish to include in an outline reliability demonstration plan the early plans for the overall evaluation of system performance.

RECOMMENDATION 4 Prior to issuing a request for proposal, the Under Secretary of Defense for Acquisition, Technology, and Logistics should mandate the preparation of an outline reliability demonstration plan that covers how the department will test a system to support and evaluate system reliability growth. The description of these tests should include the technical basis that will be used to determine the number of replications and associated test conditions and how failures are defined. The outline reliability demonstration plan should also provide the technical basis for how test and evaluation will track in a statistically defendable way the current reliability of a system in development given the likely number of government test events as part of developmental and operational testing. Prior to being included in the request for proposal for an acquisition program, the outline reliability demonstration plan should be reviewed by an expert external panel. Reliability engineers of the services involved in the acquisition in question should also have full access to the reliability demonstration plan and should be consulted prior to its finalization.

RFPs are currently based on a statement of work that contains reliability specifications for the developer and obligations for DoD. We note that the Army Materiel Systems Analysis Activity (AMSAA) has issued documents concerning language for reliability specification and contractual language for hardware and software systems that can be used as a guide for implementing the above panel recommendation.

RAISING THE PRIORITY OF RELIABILITY

A key element in improving the reliability of DoD's systems is recognizing the importance of reliability early and throughout the acquisition process. This point was emphasized in an earlier report of the Defense Science Board (2008). Many of our recommendations are consistent with the recommendations in that report (see Chapter 1). To emphasize the importance of these issues, we offer a recommendation on the need to increase the priority of reliability in the acquisition process.

At present, availability is the mandatory suitability key performance parameter, and reliability is a subordinate key system attribute. There is some evidence to suggest that when reliability falls short of its requirement, some defense acquisition personnel consider it a problem that can be addressed with more maintenance or expedited access to spare parts. Furthermore, there seems to be a belief that as long as the availability key performance parameter is met, DOT&E is likely to deem the system to be suitable. Yet DOT&E continues to find systems unsuitable because of poor reliability (see Chapter 1). This continuing deficiency supports elevation of reliability to key performance parameter status.

RECOMMENDATION 5 The Under Secretary of Defense for Acquisition, Technology, and Logistics should ensure that reliability is a key performance parameter: that is, it should be a mandatory contractual requirement in defense acquisition programs.

DESIGN FOR RELIABILITY AND RELIABILITY TESTING

As discussed throughout this report, there are two primary ways, in combination, to achieve reliability requirements in a development program: reliability can be "grown" by using a test-analyze-fix-test process, and the initial design can be developed with system reliability as an objective. The Defense Science Board's report (2008) argued that no amount of testing would compensate for deficiencies in system design due to the failure to give proper priority in the design to attain reliability requirements (see Chapter 1). We support and emphasize this conclusion.

It is important for contractors to describe the reliability management processes that they will use. Those processes should include establishment of an empowered reliability review board or team for tracking reliability from design through deployment and operations, encompassing design changes, observed failure modes, and failure and correction action analyses.

Similarly, the report of the Reliability Improvement Working Group (U.S. Department of Defense, 2008c) contained detailed advice for mandating reliability activities in acquisition contracts (see Appendix C). That report included requirements for a contractor to develop a detailed reliability model for the system that would generate reliability allocations from the system level down to lower levels, and to aggregate system-level reliability estimates based on estimates from components and subsystems. The reliability model would be updated whenever new failure modes are identified, there are updates or revisions to the failure definitions or load estimates are revised, and there are design and manufacturing changes throughout the system's life cycle. The report further called for the analysis of all failures, either in testing or in the field, until the root-cause failure mechanism is identified. In addition, the report detailed how a contractor should use a system reliability model, in conjunction with expert judgment, for all assessments and decisions about the system.

Consistent both with the report of the Reliability Improvement Working Group and with ANSI/GEIA-STD-0009, we strongly agree that proposals should provide evidence in support of the assertion that the design-for-reliability tools suggested for use and the testing schemes outlined are consistent with meeting the reliability requirement during the time allocated for development. As part of this evidence, contractors should be required to develop, and share with DoD, a system reliability model detailing how system reliability is related to that of the subsystems and components. Proposals should also acknowledge that developers will provide DoD with technical assessments, at multiple times during development, that track whether the reliability growth of the system is consistent with satisfying the reliability requirements for deployment.

We acknowledge that it is a challenge for developers to provide this information and for DoD to evaluate it before any actual development work has started. However, a careful analysis can identify proposed systems and development plans that are or are not likely to meet the reliability requirements without substantial increase in development costs and/or extensive time delays. To develop a reasonable estimate of the initial reliability corresponding to a system design, one would start with the reliability of the components and subsystems that have been used in previous systems and engineering arguments and then combine this information using reliability block diagrams, fault-tree analyses, and physics-of-failure modeling. Then, given this initial reliability, a testing plan, and the improvements that have been demonstrated by recent, related systems during development, one can roughly ascertain whether a reliability goal is feasible.

RECOMMENDATION 6 **The Under Secretary of Defense for Acquisition, Technology, and Logistics should mandate that all proposals specify the design-for-reliability techniques that the contractor will use during the design of the system for both hardware and software. The proposal budget should have a line item for the cost of design-for-reliability techniques, the associated application of reliability engineering methods, and schedule adherence.**

RECOMMENDATION 7 **The Under Secretary of Defense for Acquisition, Technology, and Logistics should mandate that all proposals include an initial plan for system reliability and qualification (including failure definitions and scoring criteria that will be used for contractual verification), as well as a description of their reliability organization and reporting structure. Once a contract is awarded, the plan should be regularly updated, presumably at major design reviews, establishing a living document that contains an up-to-date assessment of what is known by the contractor about hardware and software reliability at the component, subsystem, and system levels. The U.S. Department of Defense should have access to this plan, its updates, and all the associated data and analyses integral to their development.**

The reliability plan called for in Recommendation 7 would start with the reliability case made by DoD (see Recommendation 1). Given that the contractor is responding to an RFP with a proposal that contains an initial argument that system reliability is technically feasible, the contractor should be able to provide a more refined model that supports the assertion that the reliability requirement is achievable within the budget and time constraints of the acquisition program. As is the case of the argument provided by DoD, this should include the reliabilities of components and major subsystems along with either a reliability block diagram or a fault-tree diagram to link the estimated subsystem reliabilities to produce an estimate of the full-system reliability.

ASSESSMENT OF THE RELIABILITY OF ELECTRONIC COMPONENTS

Determining the reliability of new electronic components is a persistent problem in defense systems. Appendix D provides a critique of MIL-HDBK-217 as a method for predicting the reliability of newly developed electronic components. The basic problem with MIL-HDBK-217 is that it does not identify the root causes, failure modes, and failure mechanisms of electronic components. MIL-HDBK-217 provides predictions based on simple heuristics and regression fits to reliability data for a select number of components, as opposed to engineering design principles and physics-of-failure analyses. The result is a prediction methodology that has the following limitations: (1) the assumption of a constant failure rate is known to be false, since electronic components have instantaneous failure rates that are subject to various kinds of wear-out (due to several different types of stresses and environmental conditions) and are subject to infant mortality; (2) lack of consideration of root causes of failures, failure modes, and failure mechanisms does not allow predictions to take into consideration the load and environment history, materials, and geometries; (3) the approach taken is focused on component-level reliability prediction, therefore failing to account for manufacturing, design, system requirements, and interfaces; (4) the approach is unable to account for environmental and loading conditions in a natural way—instead they are accounted for through the use of various adjustment factors; and (5) the focus on fitting reliability data makes it impossible to provide predictions for the newest technologies and components. These limitations combine in different ways to cause the predictions from MIL-HDBK-217 to fail to accurately predict electronic component reliabilities, as has been shown by a number of careful studies, including on defense systems. Possibly most disturbingly, the use of MIL-HDBK-217 has resulted in poor ranking of the predicted reliabilities of proposed defense systems in development.

To further support this assertion, we quote from the following articles that strongly support the need to eliminate MIL-HDBK-217 in favor of a physics-of-failure approach:

- . . . it appears that this application of the Arrhenius model was not rigorously derived from physics-of-failure principles. Also, current physics-of-failure research indicates that the relationship between microelectronic device temperatures and failure rate is more complex than previously realized, necessitating explicit design consideration of temperature change, temperature rate of change, and spatial temperature gradients. And, a review of acceleration modeling theory indicates that when modeling the effect that temperature has on microelectronic device reliability, each failure mechanism should be treated separately, which is also at odds with the approach used in MIL-HDBK-217 (Cushing, 1993).
- Traditionally, a substantial amount of military and commercial reliability assessments for electronic equipment have been developed without knowledge of the root-causes of failure and the parameters which appreciably affect them. These assessments have used look-up tables from US MIL-HDBK-217 and its progeny for component failure rates which are largely based on curve fitting of field data. However, these attempts to simplify the process of reliability assessment to the point of ignoring the true mechanisms behind failure in electronics, and their life and acceleration models, have resulted in an approach which provides the design team little guidance, and may in fact harm the end-product in terms of reliability and cost. The oversimplified look-up table approach to reliability requires many invalid assumptions. One of these assumptions is that electronic components exhibit a constant failure rate. For many cases, this constant failure rate assumption can introduce a significant amount of error in decisions made for everything from product design to logistics. The constant failure rate assumption can be most detrimental when failure rates are based on past field data which includes burn-in failures, which are typically due to manufacturing defects, and/or wear out failures, which are attributed to an intrinsic failure rate which is dependent on the physical processes inherent within the component. In order to improve the current reliability assessment process, there needs to be a transition to a science-based approach for determining how component hazard rates vary as a function of time. For many applications, the notion of the constant failure rate should be replaced by a composite instantaneous hazard rate which is based on root cause failure mechanisms. A significant

amount of research has been conducted on the root-causes of failure and many failure mechanisms are well understood (Mortin et al., 1995).

- Reliability assessment of electronics has traditionally been based on empirical failure-rate models (e.g., MIL-HDBK-217) developed largely from curve fits of field-failure data. These field-failure data are often limited in terms of the number of failures in a given field environment, and determination of the actual cause of failure. Often, components are attributed incorrectly to be the cause of problems even though 30-70% of them retest OK. In MIL-HDBK-217, crucial failure details were not collected and addressed, e.g., (1) failure site, (2) failure mechanism, (3) load/environment history, (4) materials, and (5) geometries. Two consequences are: (a) MIL-HDBK-217 device failure-rate prediction methodology does not give the designer or manufacturer any insight into, or control over, the actual causes or failure since the cause-and-effect relationships impacting reliability are not captured. Yet, the failure rate obtained is often used as a reverse-engineering tool to meet reliability goals; (b) MIL-HDBK-217 does not address the design and usage parameters that greatly influence reliability, which results in an inability to tailor a MIL-HDBK-217 prediction using these key parameters. . . . A central feature of the physics-of-failure approach is that reliability modeling used for the detailed design of electronic equipment is based on root-cause failure processes or mechanisms. When reliability modeling is based on failure mechanisms, an understanding of the root-causes of failure in electronic hardware is feasible. This is because failure-mechanism models explicitly address the design parameters which have been found to influence hardware reliability strongly, including material properties, defects, and electrical, chemical, thermal, and mechanical stresses. The goal is to keep the modeling, in a particular application, as simple as feasible without losing the cause-effect relationships that advance useful corrective action. Research into physical failure mechanisms is subjected to scholarly peer review and published in the open literature. The failure mechanisms are validated through experimentation and replication by multiple researchers. Industry is now recognizing that an understanding of potential failure mechanisms leads to eliminating them cost-effectively, and is consequently demanding an approach to reliability modeling and assessment that uses knowledge of failure mechanisms to encourage robust designs and manufacturing practices (Cushing et al., 1993).

The natural focus on the part limit necessarily bounds how much attention can be given to an exhaustive consideration at the subsystems and system level—even for a physics-of-failure approach. Therefore, some judgment has to be exercised as to where to conduct detailed analyses. But if some part/component/system is determined to be "high priority," then the best available tools for addressing it should be pursued. MIL-HDBK-217 falls short in that regard.

Physics of failure has often been shown to perform better, but it does require the a priori understanding of the failure mechanisms—or the development of such. Also, MIL-HDBK-217 does not provide adequate design guidance and information regarding microelectronic failure mechanisms, and for the most part it does not include the failure rate for software, integration, manufacturing defects, etc.

Because of the limitations of MIL-HDBK-217, we want to emphasize the importance of the modern design-for-reliability techniques, particularly physics-of-failure-based methods, to support system design and reliability estimation. We wish to exclude any version of MIL-HDBK-217 from further use. Further, we realize that there is nothing particular in this regard with electronic components, and therefore, we are recommending that such techniques be used to help design for and assess the reliability of subsystems early in system development for all components in all subsystems. In particular, physics of failure should be utilized to identify potential wear-out failure modes and mitigations for enhancing long-term reliability performance.

RECOMMENDATION 8 Military system developers should use modern design-for reliability (DFR) techniques, particularly physics-of-failure (PoF)-based methods, to support system design and reliability estimation. MIL-HDBK-217 and its progeny have grave deficiencies; rather, DoD should emphasize DFR and PoF implementations when reviewing proposals and reliability program documentation.

We understand that the conversion from MIL-HDBK-217 to a new approach based on physics-of-failure modeling cannot be done overnight and that guidances, training, and specific tools need to be developed to support the change. However, this conversion can be started immediately, because the approach is fully developed in many commercial applications.

OVERSIGHT OF SOFTWARE DEVELOPMENT

If a system is software intensive or if one or more major subsystems are software intensive, then the contractor should be required to provide information on the reasons justifying the selection of the software architecture and the management plan used in code development (e.g., use of AGILE development) to produce an initial code for testing that is reasonably free of defects. Given the current lack of expertise in software engineering in the defense acquisition community, the architecture, management plan, and other specifications need to be reviewed by an outside expert panel appointed by DoD that includes users, testers, software engineers, and members from outside of the service acquiring the system. This expert panel should also review the software system design and estimates of its reliability and the uncertainty of those estimates. The panel should report to JROC, which should use this information in awarding acquisition contracts.

Software reliability growth through the test-analyze-and-fix approach can be assessed using various metrics, including build success rate, code dependency metrics, code complexity metrics, assessments of code churn and code stability, and code velocity. To assist DoD in monitoring progress toward developing reliable software, a database should be developed by the contractor to provide a constant record of an agreed-upon subset of such metrics. In addition, the contractor should maintain a sharable record of all the categories of failure and how the code was fixed in response to each discovered failure.

RECOMMENDATION 9 For the acquisition of systems and subsystems that are software intensive, the Under Secretary of Defense for Acquisition, Technology, and Logistics should ensure that all proposals specify a management plan for software development and also mandate that, starting early in development and continuing throughout development, the contractor provide the U.S. Department of Defense with full access to the software architecture, the software metrics being tracked, and an archived log of the management of system development, including all failure reports, time of their incidence, and time of their resolution.

RELIABILITY GROWTH MODELING

Reliability growth models are statistical models that link time on test, and possibly other inputs, to increases in reliability as a system proceeds through development. Because reliability growth models often fail to represent the environment employed during testing, because time on test is often not fully predictive for growth in the reliability of a system in development, and because extrapolation places severe demands on such models, they should be validated prior to use for predicting either the time at which the required reliability will be attained or for predicting the reliability attained at some point in the future. An exception to this is the use of reliability growth models early in system development, when they can help determine the scope, size, and design of the developmental testing programs.[6]

RECOMMENDATION 10 The validity of the assumptions underlying the application of reliability growth models should be carefully assessed. In cases where such validity remains in question: (1) important decisions should consider the sensitivity of results to alternative model formulations and (2) reliability growth models should not be used to forecast substantially into the future. An exception to this is early in system development, when reliability growth models, incorporating relevant historical data, can be invoked to help scope the size and design of the developmental testing programs.

When using reliability growth models to scope developmental testing programs, there are no directly relevant data to validate modeling assumptions. Historical reliability growth patterns experienced for similar classes of systems can be reviewed, however. These should permit the viability of the proposed reliability growth trajectory for the subject system to be assessed. They should also support the allocation of adequate budget reserves that may become necessary if the originally envisioned reliability growth plan turns out to be optimistic.

RELIABILITY GROWTH TESTING

One can certainly argue that one reason that many defense systems fail to achieve their reliability requirements is because there are too many defects that have not been discovered when the system enters operational testing. Given the limitations in discovering reliability defects in both developmental and operational tests, most of the effort to find reliability problems prior to fielding needs to be assumed by the contractor. Although a great deal can be done at the design stage, using design-for-reliability techniques, some additional reliability growth will need to take place through testing and fixing the reliability problems that are discovered, and the majority of the reliability growth through testing has to take place through contractor testing. Consequently, DoD has an interest in monitoring the testing that is budgeted for in acquisition proposals and in monitoring the resulting progress toward the system's reliability requirements.

Because the contractor has control of the only direct test information

[6]Elements of Recommendations 7, 9, and 10, which concern plans for design-for-reliability and reliability testing for both hardware and software systems and subsystems, are sometimes referred to as a "reliability case": for details, see Jones et al. (2004).

on the reliability of both subsystems and the full system through early development, granting DoD access to such information can help DoD monitor progress on system development and the extent to which a system is or is not likely to satisfy its reliability requirements. In addition, such access can enable DoD to select test designs for developmental and operational testing to verify that design faults have been removed and so that relatively untested components and subsystems are more thoroughly tested.

Thus, it is critical that DoD be provided with information both on reliability test design at the proposal stage to examine whether such plans are sufficient to support the necessary degree of reliability growth, and on reliability test results during development to enable the monitoring of progress toward attainment of requirements. The information on reliability test design should include the experimental designs and the scenario descriptions of such tests, along with the resulting test data, for both the full system and all subsystem assemblies, as well as the code for and results of any modeling and simulation software that were used to assess reliability. The information should cover all types of hardware testing, including testing under operationally relevant conditions, and any use of accelerated or highly accelerated testing.[7] The contractor should also provide DoD with information on all types of software testing, including the results of code reviews, automated testing, fault seeding, security testing, and unit test coverage.

In order to ensure that this information is provided to DoD, acquisition contracts will need to be written so that this access is mandated and proposals will need to state that contractors agree to share this information. This information sharing should occur at least at all design reviews throughout system development. This sharing of information should enable DoD to assess system reliability at the time of the delivery of system prototypes, which can help DoD make better decisions about whether to accept delivery of prototypes.

> **RECOMMENDATION 11 The Under Secretary of Defense for Acquisition, Technology, and Logistics should mandate that all proposals obligate the contractor to specify an initial reliability growth plan and the outline of a testing program to support it, while recognizing that both of these constructs are preliminary and will be modified through development. The required plan will include, at a minimum, information on whether each test is a test of components, of subsystems, or of the full system; the scheduled dates; the test design; the test scenario conditions; and the number of replications in each scenario. If a test is an accelerated**

[7]For further details on the information that should be provided to DoD, see National Research Council (2004).

test, then the acceleration factors need to be described. The contractor's budget and master schedules should be required to contain line items for the cost and time of the specified testing program.

RECOMMENDATION 12 The Under Secretary of Defense for Acquisition, Technology, and Logistics should mandate that contractors archive and deliver to the U.S. Department of Defense, including to the relevant operational test agencies, all data from reliability testing and other analyses relevant to reliability (e.g., modeling and simulation) that are conducted. This should be comprehensive and include data from all relevant assessments, including the frequency under which components fail quality tests at any point in the production process, the frequency of defects from screenings, the frequency of defects from functional testing, and failures in which a root-cause analysis was unsuccessful (e.g., the frequency of instances of failure to duplicate, no fault found, retest OK). It should also include all failure reports, times of failure occurrence, and times of failure resolution. The budget for acquisition contracts should include a line item to provide DoD with full access to such data and other analyses.

MODELING IN CONJUNCTION WITH ACCELERATED TESTING

Similar to the panel's concerns above about the use of reliability growth models for extrapolation, models used in conjunction with accelerated testing linking extreme use to normal use also use extrapolation and therefore need to be validated for this use. The designs of such tests are potentially complicated and would therefore also benefit from a formal review. Such validation and formal review are particularly important when accelerated testing inference is of more than peripheral importance, for example, if applied at the major subsystem or system level, and there is inadequate corroboration provided by limited system testing and the results are central to decision making on system promotion.

RECOMMENDATION 13 The Office of the Secretary of Defense for Acquisition, Technology, and Logistics, or, when appropriate, the relevant service program executive office, should enlist independent external, expert panels to review (1) proposed designs of developmental test plans critically reliant on accelerated life testing or accelerated degradation testing and (2) the results and interpretations of such testing. Such reviews should be undertaken when accelerated testing inference is of more than peripheral importance—for example, if applied at the major subsystem or system level, there is inadequate corroboration

provided by limited system testing, and the results are central to decision making on system promotion.

Software systems present particular challenges for defense acquisition. Complicated software subsystems and systems are unlikely to be comprehensively tested in government developmental or operational testing because of the current lack of software engineers at DoD. Therefore, such systems should not be accepted for delivery from a contractor until the contractor has provided sufficient information for an assessment of their readiness for use. To provide for some independent testing, the contractor should provide DoD with fully documented software that conducts automated software testing for all of its software-intensive subsystems and for the full system when the full system is a software system. This documentation will enable DoD to test the software for many times the order of magnitude of replications that would otherwise be possible in either developmental or operational testing.

RECOMMENDATION 14 **For all software systems and subsystems, the Under Secretary of Defense for Acquisition, Technology, and Logistics should mandate that the contractor provide the U.S. Department of Defense (DoD) with access to automated software testing capabilities to enable DoD to conduct its own automated testing of software systems and subsystems.**

DESIGN CHANGES

Changes in design during the development of a system can have significant effects on the system's reliability. Consequently, developers should be required to include descriptions of the impact of substantial system design changes and how such changes required the modification of plans for design-for-reliability activities and plans for reliability testing. Any changes in fund allocation for such activities should be communicated to DoD. This information will help to support more efficient DoD oversight of plans for design for reliability and reliability testing.

RECOMMENDATION 15 **The Under Secretary of Defense for Acquisition, Technology, and Logistics should mandate the assessment of the impact of any major changes to system design on the existing plans for design-for-reliability activities and plans for reliability testing. Any related proposed changes in fund allocation for such activities should also be provided to the U.S. Department of Defense.**

INFORMATION ON OPERATIONAL ENVIRONMENTS

Inadequate communication between the prime contractor and subcontractors can be a source of difficulties in developing a reliable defense system. In particular, subcontractors need to be aware of the stresses and strains, loads, and other sources of degradation that the components they supply will face. Therefore, acquisition contracts need to include the contractor's plan to ensure the reliability of components and subsystems, especially those that are produced by subcontractors and those that are commercial off-the-shelf systems. For off-the-shelf systems, the risks associated with using a system in an operational environment that differs from its intended environment should be assessed. To do so, the government has to communicate the operational environment to the contractor, and the contractor, in turn, has to communicate that information to any subcontractors.

RECOMMENDATION 16 The Under Secretary of Defense for Acquisition, Technology, and Logistics should mandate that contractors specify to their subcontractors the range of anticipated environmental load conditions that components need to withstand.

RECOMMENDATION 17 The Under Secretary of Defense for Acquisition, Technology, and Logistics should ensure that there is a line item in all acquisition budgets for oversight of subcontractors' compliance with reliability requirements and that such oversight plans are included in all proposals.

ACQUISITION CONTRACTS

The above recommendations would require contractors to lay out their intended design for reliability and reliability testing activities in acquisition proposals. The level of effort should be a factor in awarding acquisition contracts. In addition, to ensure that the general level of effort for reliability is sufficient, contractors should provide to DoD their budgets for these activities, and those budgets should be protected, even in the face of unanticipated problems.

RECOMMENDATION 18 The Under Secretary of Defense for Acquisition, Technology, and Logistics should mandate that proposals for acquisition contracts include appropriate funding for design-for-reliability activities and for contractor testing in support of reliability growth. It should be made clear that the awarding of contracts will include consideration of such fund allocations. Any changes to such allocations after a contract award should consider the impact on probability of mission

success and on life-cycle costs, and at the minimum, require approval at the level of the service component acquisition authority.

DELIVERY OF PROTOTYPES FOR DEVELOPMENTAL TESTING

We argue throughout this report that both developmental and, especially, operational testing as currently practiced are limited in their ability to discover reliability problems in defense system prototypes. We recommend ways in which government testing can be made more effective in identifying reliability problems, for instance, by adding aspects of operational realism to developmental testing. Also, by targeting DoD developmental testing to those components that were problematic in development, developmental testing can be made more productive. And, for software, by acquiring capabilities for software testing from the contractor, DoD can play the role of an independent software tester.

However, even after implementing these recommendations, it is likely that most reliability growth through testing will need to be achieved by contractor testing, rather than through DoD's developmental or operational testing. Furthermore, even though there will likely be appreciable reliability growth as a result of developmental and operational testing, not only is it limited by the lack of operational realism in developmental testing and the short time frame of operational testing, but there also will likely be reliability "decline" due to the developmental test/operational test gap (see Chapter 8.) Although the magnitudes of these increases and decreases cannot be determined a priori, one can increase overall reliability of all systems by requiring that prototypes achieve their reliability requirements on delivery to DoD.

RECOMMENDATION 19 The Under Secretary of Defense for Acquisition, Technology, and Logistics should mandate that prior to delivery of prototypes to the U.S. Department of Defense for developmental testing, the contractor must provide test data supporting a statistically valid estimate of system reliability that is consistent with the operational reliability requirement. The necessity for this should be included in all requests for proposals.

This recommendation should not preclude the early delivery of subsystems that are considered final to DoD, while development work continues on other parts of the system, when doing so is considered beneficial by both the contractor and by DoD.

The estimation of system reliability called for in this recommendation would likely need to combine information from full-system testing done late in development with component- and subsystem-level testing done earlier, and it could also use estimates from previous versions of the system at either the full-system, subsystem, or component levels. In this way, the contractor will be able to justify delivery of prototypes. Such assessments will, at times, be later demonstrated by DoD to be high or low, but this approach will support a learning process about how to better merge such information in the future.

DEVELOPMENTAL TESTING

The monitoring of system reliability prior to operational testing is important, because it is likely that relying on operational tests to expose reliability problems will result in too many defense systems exhibiting deficient system reliability when fielded (see Chapter 7). Yet accounting for the differences in conditions between developmental and operational testing remains a challenge. One possible approach to meet this challenge is, to the extent possible, make greater use of test conditions in nonaccelerated developmental testing that reflect operational conditions. Then, DoD must somehow estimate what the system reliability will likely be under operational conditions based on results from developmental testing.

Schedule pressures, availability of test facilities, and testing constraints necessarily limit the capability of contractors to consistently be able to carry out testing under circumstances that mimic operational use. It thus remains important for DoD to provide its own assessment of a system's operationally relevant levels prior to making the decision to proceed to operational testing. This assessment is best accomplished through the use of a full-system test in environments that are as representative of actual use as possible.

RECOMMENDATION 20 Near the end of developmental testing, the Under Secretary of Defense for Acquisition, Technology, and Logistics should mandate the use of a full-system, operationally relevant developmental test during which the reliability performance of the system will equal or exceed the required levels. If such performance is not achieved, then justification should be required to support promotion of the system to operational testing.

OPERATIONAL TESTING

Operational testing provides an assessment of system reliability in as close to operational circumstances as possible. As such, operational testing provides the best indication as to whether a system will meet its reliability requirement when fielded. Failure to meet the reliability requirement during operational test is a serious deficiency for a system and should generally be the cause for delaying promotion of the system to full-rate production until modifications to the system design can be made to improve system reliability to meet the requirement.

RECOMMENDATION 21 The U.S. Department of Defense should not pass a system that has deficient reliability to the field without a formal review of the resulting impacts the deficient reliability will have on the probability of mission success and system life-cycle costs.

Reliability deficiencies can continue to arise after deployment, partly because however realistic an operational test strives to be, it will always differ from field deployment. Field operations can stress systems in unforeseen ways and reveal failure modes that were not likely to have been unearthed in either developmental or operational testing. In addition, feedback and system retrofits from field use can further improve reliability of a given system and can also improve the reliability of subsequent related systems if lessons are learned and communicated. Therefore, the support and enhancement of such feedback loops should be a DoD priority. One way to do so is through continuous monitoring of reliability performance in fielded systems.

RECOMMENDATION 22 The Under Secretary of Defense for Acquisition, Technology, and Logistics should emplace acquisition policies and programs that direct the services to provide for the collection and analysis of postdeployment reliability data for all fielded systems, and to make that data available to support contractor closed-loop failure mitigation processes. The collection and analysis of such data should be required to include defined, specific feedback about reliability problems surfaced in the field in relation to manufacturing quality controls and indicate measures taken to respond to such reliability problems. In addition, the contractor should be required to implement a comprehensive failure reporting, analysis and corrective action system that encompasses all failures (regardless whether failed items are restored/repaired/replaced by a different party, e.g., subcontractor or original equipment manufacturer).

Problems can arise when a contractor changes its subcontractors or suppliers. If this is done without proper oversight, then it can result in substantial reductions in reliability. Therefore, contractors should be required to document the reason for such changes and estimate the likelihood of mission success and modified life-cycle costs due to such changes in the fielded system. The document detailing the implications of such changes should be reviewed by an external panel of reliability and system experts. If the review finds that there is the potential for a substantial decrease in system reliability, then USD (AT&L) should not approve the change.

> **RECOMMENDATION 23 After a system is in production, changes in component suppliers or any substantial changes in manufacturing and assembly, storage, shipping and handling, operation, maintenance, and repair should not be undertaken without appropriate review and approval. Reviews should be conducted by external expert panels and should focus on impact on system reliability. Approval authority should reside with the program executive office or the program manager, as determined by the U.S. Department of Defense. Approval for any proposed change should be contingent upon certification that the change will not have a substantial negative impact on system reliability or a formal waiver explicitly documenting justification for such a change.**

This report is focused on activities that are undertaken prior to the end of operational testing. However, approaches to manufacturing and assembly, storage, shipping and handling, operation, maintenance, and repair also affect system reliability. In particular, it is crucial that supply-chain participants have the capability to produce the parts and materials of sufficient quality to support meeting a system's final reliability objectives. Because of changes in technology trends, the evolution of complex supply-chain interactions, a cost-effective and efficient parts selection and management process is needed to perform this assessment.

INTERMEDIATE RELIABILITY GOALS

Setting target values for tracking system reliability during development is important for discriminating between systems that are likely to achieve their reliability requirements and those that will struggle to do so. Through early identification of systems that are having problems achieving the required reliability, increased emphasis and resources can be placed on design for reliability or reliability testing, which will often provide a remedy. Given the difficulty of modifying systems later in development, it is critical that such problems are identified as early in the process as possible.

Target reliability values at specified times could be set both prior to or after delivery of prototypes from the contractor to DoD. Prior to delivery of prototypes for developmental testing, intermediate target values could be set by first determining the initial reliability level, based only on design-for-reliability activities, prior to most subsystem- or system-level testing. Then the contractor, possibly jointly with DoD, would decide what model of reliability as a function of time should be used to link this initial level of reliability with the reliability requirement to support delivery of prototypes to DoD (see Chapter 4). Such a function could then be used to set intermediate reliability targets.

As noted throughout the report, the number of test replications carried out by a contractor is likely to be very small for any specific design configuration. Therefore, such estimates are likely to have large variances. This limitation needs to be kept in mind in setting up decision rules that aim to identify systems that are unlikely to improve sufficiently to make their reliability requirement absent additional effort (see Chapter 7).

After prototype delivery, the specified initial reliability level could be the reliability assessed by the contractor on delivery or during early full-system developmental testing; the final level would be the specified requirement, and its date would be the scheduled time for initiation of operational testing. Again, the contractor and DoD would have to decide what function should be used to link the initial level of reliability with the final value and the associated dates used to fit target values. Having decided on that, intermediate reliability targets can be easily determined. As noted above, the variances of such reliability estimates would need to be considered in any decision rules pertaining to whether a system is or is not ready to enter into operational testing.

In each of these applications, one is merely fitting a curve of hypothesized reliabilities over time that will associate the initial reliability to the reliability goal over the specified time frame. One can imagine curves in which most of the change happens early in the time frame, other curves with relatively consistent changes over time, and myriad other shapes. Whatever curve is selected, it is this curve that will be used to provide intermediate reliability targets to compare with the current estimates of reliability, with the goal of using discrepancies from the curve to identify systems that are unlikely to meet their reliability requirement in the time allotted. Experience with similar systems should provide information about the adequacy of the length, number, and type of test events to achieve the target reliability. Clearly, the comparisons to be made between the estimated system reliability, its estimated standard error, and the target values are most likely to occur at the time of major developmental (and related) test events or during major system reviews.

With respect to the second setting of target values, the appropriate time to designate target values for reliability is after delivery of prototypes because reliability levels cannot be expected to appreciably improve as a result of design flaws discovered during operational testing. As noted throughout the report, operational testing is generally focused on identifying deficiencies in effectiveness, not in suitability, and fixing flaws discovered at this stage is both expensive and risky (see Chapter 8). Unfortunately, late-stage full-system developmental testing, as currently carried out, may also be somewhat limited in its potential to uncover flaws in reliability design due to its failure to represent many aspects of operational use (see Chapter 8; also see National Research Council, 1998). As the Defense Science Board emphasized (U.S. Department of Defense, 2008a), testing cannot overcome a poor initial design. Therefore, it is important to insist that more be done to achieve reliability levels at the design stage, and therefore, the goals for initial reliability levels prior to entry into developmental testing should be set higher than are presently the case. Starting out on a growth trajectory by having an initial design that provides too low an initial reliability is a major reason that many systems fail to attain their requirement. Unfortunately, one cannot, a priori, provide a fixed rule for an initial system reliability level in order to have a reasonable chance of achieving the reliability requirement prior to delivery of prototypes or prior to operational testing. At the very least, one would expect that such rules would be specific to the type of system.

More generally, several key questions remain unanswered concerning which design-for-reliability techniques and reliability growth tests are most effective for which types of defense systems, the order in which they are most usefully applied, and the total level of effort that is needed for either design for reliability or for reliability growth.

To help clarify these and other important issues, DoD should collect and archive, for all recent acquisition category I systems (see Chapter 1), the estimated reliability for at least five stages of development:

1. the level achieved by design alone, prior to any contractor testing,
2. the level at delivery of prototypes to DoD,
3. the level at the first system-level government testing,
4. the level achieved prior to entry into operational testing, and
5. the level assessed at the end of operational testing.

Analyses of the data would provide information as to the degree of improvement toward reliability requirements that are feasible for different types of systems at each stage of the development process. Such an analysis could be useful input toward the development of rules as to what levels of reliability should be evidence of promotion to subsequent stages of development.(Such an analysis would require some type of taxonomy of defense systems in which the patterns of the progression to requirements were fairly comparable for all the members in a cell.)

Analysis of these data may also determine the factors that play a role in achieving reliability requirements. For example, it would be of great importance to determine which design-for-reliability techniques or what budgets for design for reliability were predictive of higher or lower rates of initial full system developmental test reliability levels. Similarly, it would also be important to determine what testing efforts, including budgets for reliability testing, and what kinds of tests used, were successful in promoting reliability growth. One could also consider the achieved reliabilities of related systems as predictors.

There are likely to be considerable additional benefits if DoD sets up a database with these and other variables viewed as having a potential impact on reliability growth and reliability attainment from recent past and current acquisition programs. For example, this database could also be useful in the preparation of cost-benefit analyses and business case analyses to support the conduct of specific reliability design tasks and tests. Such kinds of databases are commonplace for the best performing commercial system development companies, because they support the investigation of the factors that are and are not related to the acquisition of reliable systems. While it is more difficult for defense systems, any information on the reliability of fielded systems could also be added to such a database.

RECOMMENDATION 24 The Under Secretary of Defense for Acquisition, Technology, and Logistics should create a database that includes three elements obtained from the program manager prior to government testing and from the operational test agencies when formal developmental and operational tests are conducted: (1) outputs, defined as the reliability levels attained at various stages of development; (2) inputs, defined as the variables that describe the system and the testing conditions; and (3) the system development processes used, that is, the reliability design and reliability testing specifics. The collection of these data should be carried out separately for major subsystems, especially software subsystems.

Analyses of these data should be used to help discriminate in the future between development programs that are and are not likely to attain reliability requirements. Such a database could also profitably include information on the reliability performance of fielded systems to provide a better "true" value for reliability attainment. DoD should seek to find techniques by which researchers can use extracts from this database while protecting against disclosure of proprietary and classified information. Finally, DoD should seek to identify leading examples of good practice of the development of reliable systems of various distinct types and collect them in a casebook for use by program managers.

Once it is better understood how near the initial reliability needs to be to the required level to have a good chance of attaining the required level prior to entry into operational testing, acquisition contracts could indicate the reliability levels that need to be attained by the contractor before a system is promoted to various stages of development. (Certainly, when considering whether demonstrated reliability is consistent with targets from a reliability growth curve, the impact of any impending corrective actions should be factored into such assessments.)

OVERSIGHT AND RESEARCH

We believe that collectively, the above recommendations will address what may have been a not uncommon practice of contractors' submitting proposals that simply promised to produce a highly reliable defense system without providing details regarding the measures that would be taken to ensure this. Proposals were not obligated to specify which, if any, design-for-reliability methodologies would be used to achieve as high an initial reliability as possible prior to formal testing, and proposals were not obligated to specify the number, size, and types of testing events that would be carried out to "grow" reliability from its initial level to the system's required level. Contractors were also not required to provide the associated budgets or impacts on schedule of delivery of prototypes.

Partly as a result of the absence of these details, there was no guarantee that reliability-related activities would take place. In fact, proposals that did allocate a substantial budget and development time and detail in support of specific design-for-reliability procedures and comprehensive testing have been implicitly penalized because their costs of development were higher and their delivery schedules were longer in comparison with proposals that made less specific assertions as to how their reliability requirements would be met. Our recommendations above will level the playing field by removing any incentive to reduce expenditures on reliability growth or reliability testing to lower a proposal's cost and so increase the chances of winning the contract.

Systems should have objective reliability thresholds that will serve as "go/no–go" gates that are strictly enforced, preventing promotion to the next stage of development or to the field unless those thresholds have been attained. At each of the decision points for development of a system, if the assessed level of reliability is considerably different from the reliability growth curve, then the system should not be promoted to the next level unless there is a compelling reason to do so.

A number of the above recommendations either explicitly (by mentioning an expert external panel) or implicitly utilize expertise in reliability and associated methods and models. In our opinion, DoD currently does not have sufficient expertise in reliability to provide satisfactory oversight of the many ACAT I acquisition programs. Therefore, we recommend that DoD initiate steps to acquire additional expertise.

RECOMMENDATION 25 To help provide technical oversight regarding the reliability of defense systems in development, specifically, to help

develop reliability requirements, to review acquisition proposals and contracts regarding system reliability, and to monitor acquisition programs through development, involving the use of design-for-reliability methods and reliability testing, the U.S. Department of Defense should acquire, through in-house hiring, through consulting or contractual agreements, or by providing additional training to existing personnel, greater access to expertise in these five areas: (1) reliability engineering, (2) software reliability engineering, (3) reliability modeling, (4) accelerated testing, and (5) the reliability of electronic components.

Lastly, our statement of task asked the panel to explore ways in which reliability growth processes and various models could be used to improve the development and performance of defense systems. In its work to produce this report, the panel did identify a number of research areas that DoD might consider supporting in the future: reliability design support, comprehensive reliability assessment, and assessing aspects of reliability that are difficult to observe in development and operational testing.

With regard to reliability design support, research on the relationship between physical failure mechanisms and new technologies seems warranted. We note three difficult issues in this area:

- assessment of system reliability that is complicated by interdependence of component functionality and tolerances; the nonlinear nature of fault development and expression; and the variation of loading conditions, maintenance activities, and operational states;
- the relationship between reliability assessment during system development in comparison to full-rate manufacturing; and
- assessment of the impact of high-frequency signals across connections in a system.

There is much that could be gained from research on assessment methodologies. We suggest, in particular:

- reliability assessment of advanced commercial electronics that addresses the next generation high-density semi-conductors and nano-scale electronic structures; copper wirebonds; environmentally friendly molding compounds; advanced environmentally friendly consumer materials; and power modules for vehicle-aerospace applications and batteries; and
- modeling the inherent uncertainty due to variation in supply and manufacturing chains in an approach similar to reliability block diagrams for the purpose of reliability prediction; and creation of traditional reliability metrics from physics-of-failure models.

There are also intriguing near- and long-term reliability issues that would be difficult to observe in development and operational testing. We note, particularly:

- the identification, characterization, and modeling of the effects of defects that can lead to early failures (infant mortality);
- reliability qualification for very long life cycles containing simultaneous impacts of multiple, combined types of stresses;
- built-in self-diagnosis of sensor degradation as systems are being increasingly instrumented with sensing functions and degradation of the sensors can lead to erroneous operation in that degradation goes undetected; and
- long-term (e.g., space flight, storage) failure models and test methods.

References

Alam, M., Azarian, M., and Pecht, M. (2012). Reliability of embedded planar capacitors with epoxy-BaTiO3 composite dielectric during temperature-humidity-bias tests. *IEEE Transactions on Device and Materials*, *12*(1), 86-93.

Ascher, H. (1968). Evaluation of repairable system reliability using the 'bad-as-old' concept. *IEEE Transactions on Reliability*, *R-17*(2), 105-110.

Azarian, M., Keimasi, M., and Pecht, M. (2006). *Non-Destructive Techniques for Detection of Defects in Multilayer Ceramic Capacitors* (pp. 125-130). Paper presented at Components for Military and Space Electronics Conference, February 6-9, Los Angeles, CA.

Basili, V., Briand, L., and Melo, W. (1996). A validation of object oriented design metrics as quality indicators. *IEEE Transactions on Software Engineering*, *22*(10), 751-761.

Bastani, F.B., and Ramamoorthy, C.V. (1986). Input-domain-based models for estimating the correctness of process control programs. In A. Serra, and R.E. Barlow (Eds.), *Reliability Theory* (pp. 321-378). Amsterdam: North-Holland.

Biyani, S., and Santhanam, P. (1998). Exploring defect data from development and customer usage on software modules over multiple releases. *Proceedings of International Symposium on Software Reliability Engineering*, 316-320.

Blischke, W.R., and Prabhakar Murthy, D.N. (2000). *Reliability: Modeling, Prediction, and Optimization.* New York: John Wiley & Sons.

Box, G.E.P., Hunter, J.S., and Hunter, W.G. (2005). *Statistics for Experimenters: Design, Innovation, and Discovery, Second Edition.* Hoboken, NJ: John Wiley & Sons.

Boydston, A., and Lewis, W. (2009). *Qualification and Reliability of Complex Electronic Rotorcraft Systems.* Paper presented at the AHS Specialists Meeting on Systems Engineering, October 15-16, Harbor, CT.

Briand, L.C., Wuest, J., Ikonomovski, S., and Lounis, H. (1999). Investigating quality factors in object-oriented designs: An industrial case study. *Proceedings of International Conference on Software Engineering* (pp. 345-354).

Cheng, S., Tom, K., Thomas, L., and Pecht, M. (2010a). A wireless sensor system for prognostics and health management. *IEEE Sensors Journal*, *10*(4), 856-862.

Cheng, S., Azarian, M., and Pecht, M. (2010b). Sensor systems for prognostics and health management. *Sensors (Basel, Switzerland), 10*(6), 5774-5797.

Chidamber, S.R., and Kemerer, C.F. (1994). A metrics suite for object oriented design. *IEEE Transactions on Software Engineering, 20*(6), 476-493.

Chillarege, R., Kao, W-L., and Condit, R.G. (1991). Defect type and its impact on the growth curve. *Proceedings of the International Conference on Software Engineering*, 246-255.

Collins, D.H., and Huzurbazar, A.V. (2012). Prognostic models based on statistical flowgraphs. *Applied Stochastic Models in Business and Industry, 28*(2), 141-151.

Crow, L.H. (1974). *Reliability Analysis for Complex Repairable Systems.* Technical Report TR138. Aberdeen Proving Ground, MD: U.S. Army Materiel Systems Analysis Activity.

Crow, L.H. (1983). Reliability growth projection from delayed fixes. *Proceedings of the 1983 Annual Reliability and Maintainability Symposium* (pp. 84-89).

Crow, L.H. (2008). A methodology for managing reliability growth during operational mission profile testing. *Proceedings of the 2008 Annual Reliability and Maintainability Symposium* (pp. 48-53).

Crow, L.H., and Singpurwalla, N.D. (1984). An empirically developed Fourier series model for describing software failures. *IEEE Transactions on Reliability, 33*(2), 176-183.

Crowder, M.J., Kimber, A.C, Smith, R.L, and Sweeting, T.J. (1991). *Statistical Analysis of Reliability Data.* London: Chapman & Hall.

Cushing, M.J. (1993). Another perspective on the temperature dependence of microelectronic-device reliability. *Proceedings of the 1993 Annual Reliability and Maintainability Symposium* (pp. 333-338).

Cushing, M.J. (2012). *ATEC Reliability Growth Case Studies and Lessons Learned.* Available: https://secure.inl.gov/isrcs2012/Presentations/ScienceOfTest_Cushing.pdf.

Cushing, M.J., Mortin, D.E., Stadterman, T.J., and Malhotra, A. (1993). Comparison of electronics reliability assessment approaches. *IEEE Transactions on Reliability, 42*(4), 542-546.

Dalton, K., and Hall, J.B. (2010). *Implementing New RAM Initiatives in Army Test and Evaluation.* Annual Reliability and Maintainability Symposium, San Jose, CA. Available: http://ieeexplore.ieee.org/stamp/stamp.jsp?tp=&arnumber=5448051 [October 2014].

Das, D. (2012). *Prognostics and Health Management: Utilizing the Life Cycle Knowledge to Reduce Life Cycle Cost.* Prepared for the First International Symposium on Physics and Technology of Sensors, March 7-10, Pune, India. Available: http://ieeexplore.ieee.org/xpls/abs_all.jsp?arnumber=6260908 [October 2014].

Denaro, G., Morasca, S., and Pezze, M. (2002). *Deriving Models of Software Fault-Proneness.* Prepared for the 14th International Conference on Software Engineering and Knowledge Engineering, New York. Available: http://dl.acm.org/citation.cfm?id=568824 [October 2014].

Drake, J.E. (1987). *Discrete Reliability Growth Models Using Failure Discounting* (Unpublished master's thesis). U.S. Naval Postgraduate School, Monterey, CA.

Duane, J.T. (1964). Learning curve approach to reliability monitoring. *IEEE Transactions on Aerospace and Electronic Systems, 2*(2), 563-566.

Duran, J.W., and Wiorkowski, J.J. (1981). Capture-recapture sampling for estimating software error content. *IEEE Transactions on Software Engineering, 7*(1), 147-148.

Ehrenberger, W. (1985). Statistical testing of real-time software. In W.J. Quirk (Ed.), *Verification and Validation of Real-Time Software* (pp. 147-178). New York: Springer-Verlag.

El-Emam, K. (2000). *A Methodology for Validating Software Product Metrics.* NRC/ERC-1076. Ottawa, Canada: National Research Council of Canada.

Ellner, P.M., and Hall, J.B. (2006). *AMSAA Planning Model Based on Projection Methodology (PM2).* Technical Report 2006-9. Aberdeen Proving Ground, MD: U.S. Army Materiel Systems Analysis Activity.

Ellner, P.M., and Trapnell, P.B. (1990). *AMSAA Reliability Growth Data Study.* Interim Note IN-R-184. Aberdeen Proving Ground, MD: U.S. Army Materiel Systems Analysis Activity.

Elsayed, E.A. (2012). Overview of reliability testing. *IEEE Transactions on Reliability*, *61*(2), 282-291.

Escobar, L.A., Meeker, W.Q., Kugler, D.-L., and Kramer, L.L. (2003). Accelerated destructive degradation tests: Data, models, and analysis. *Mathematical Methods in Reliability,* 319-338.

Foucher, B., Boullie, J., Meslet, B., and Das, D. (2002). A review of reliability prediction methods for electronic devices. *Microelectronics Reliability, 42(*8), 1155-1162.

Fries, A., and Sen, A. (1996). A survey of discrete reliability growth models. *IEEE Transactions on Reliability*, *R-45*(4), 582-604.

George, E., Das, D., Osterman, M., and Pecht, M. (2009). *Physics of Failure Based Virtual Testing of Communication Hardware.* Prepared for ASME International Mechanical Engineering Congress and Exposition, November 13-19, Lake Buena Vista, FL.

Goel, A., and Okumoto, K. (1979). Time-dependant error-detection rate model for software reliability and other performance measures. *IEEE Transactions on Reliability*, *28*(3), 206-211.

Graves, T.L., Karr, A.F., Marron, J.S., and Siy, H. (2000). Predicting fault incidence using software change history. *IEEE Transactions in Software Engineering*, *26*(7), 653-661.

Gyimóthy, T., Ferenc, R., and Siket, I. (2005). Empirical validation of object-oriented metrics on open source software for fault prediction. *IEEE Transactions in Software Engineering*, *31*(10), 897-910.

Hall, J.B., Ellner, P.M., and Mosleh, A. (2010). Reliability growth management metrics and statistical methods for discrete-use systems. *Technometrics, 52*(4), 379-389.

Hamada, M.S., Wilson, A.G., Reese, C.S., and Martz, H.F. (2008). *Bayesian Reliability.* New York: Springer.

Han, J., and Kamber, M. (2006). *Data Mining: Concepts and Techniques Second Edition.* San Francisco, CA: Morgan Kaufmann.

Hobbs, G.K. (2000). *Accelerated Reliability Engineering: HALT and HASS.* Hoboken, NJ: John Wiley & Sons.

Information Technology Association of America. (2008). *Reliability Program Standard for Systems Design, Development, and Manufacturing: GEIA-STD-0009.* Arlington, VA: Author.

Jaai, R., and Pecht, M. (2010). A prognostics and health management roadmap for information and electronics-rich systems. *Microelectronics Reliability, 50*, 317-323.

Jacoby, R., and Masuzawa, K. (1992). *Test Coverage Dependent Software Reliability Estimation by the HGD Model.* Prepared for Third International Symposium on Software Reliability Engineering, October 7-10, Research Triangle Park, NC.

Jelinksi, Z., and Moranda, P.B. (1972). Software reliability research. In W. Freiberger (Ed.), *Statistical Computer Performance Evaluation* (pp. 465-497). New York: Academic Press.

Jones, J.A., Marshall J., and Newman R. (2004). *The Reliability Case in the REMM Methodology.* Prepared for the Annual Reliability and Maintainability Symposium, January 26-29.

Jones, W., and Vouk, M.A. (1996). Software reliability field data analysis. In M. Lyu (Ed.), *Handbook of Software Reliability Engineering* (Chapter 11). New York: McGraw-Hill.

Keimasi, M., Ganesan, S., and Pecht, M. (2006). Low temperature electrical measurements of silicon bipolar Monolithic Microwave Integrated Circuit (MMIC) amplifiers. *Microelectronics Reliability*, *46*(2-4), 326-334.

Khoshgoftaar, T.M., Allen, E.B., Goel, N., Nandi, A., and McMullan, J. (1996). Detection of software modules with high debug code churn in a very large legacy system. *Proceedings of International Symposium on Software Reliability Engineering*, 364-371.

Krasich, M. (2009). How to estimate and use MTTF/MTBF—Would the real MTBF please stand up? *Proceedings of the 2009 Annual Reliability and Maintainability Symposium* (pp. 353-359).

Kumar, S., Vichare, N.M., Dolev, E., and Pecht, M. (2012). A health indicator method for degradation detection of electronic products. *Microelectronics Reliability, 52*(2), 439-445.

Li, M., and Meeker, W.Q. (2014). Application of Bayesian methods in reliability data analysis. *Journal of Quality Technology, 46*(1), January. Available: http://asq.org/pub/jqt/past/vol46-issue1/index.html [October 2014].

Littlewood, B., and Verrall, J.L. (1973). A Bayesian reliability growth model for computer software. *Applied Statistics, 22*, 332-346.

Liu, G. (1987). A Bayesian assessing method of software reliability growth. In S. Osaki and J. Cao (Eds.), *Reliability Theory and Applications* (pp. 237-244). Singapore: World Scientific.

Lloyd, D.K. (1987). *Forecasting Reliability Growth.* Prepared for the 33rd Annual Technical Meeting of the Institute of Environmental Science, May 5-7, San Jose, CA.

Long, E.A, Forbes, J., Hees, J., and Stouffer, V. (2007, June 1). *Empirical Relationships Between Reliability Investments and Life-Cycle Support Costs.* Report SA701T1. McLean, VA: LMI Government Consulting.

Mathew, S., Das, D., Osterman, M., Pecht, M., Ferebee, R., and Clayton, J. (2007). Virtual remaining life assessment of electronic hardware subjected to shock and random vibration life cycle loads. *Journal of the IEST, 50*(1), 86-97.

Mathew, S., Alam, M., and Pecht, M. (2012). Identification of failure mechanisms to enhance prognostic outcomes. *ASM Journal of Failure Analysis and Prevention, 12*(1), 66-73.

McCabe, T.J. (1976). A complexity measure. *IEEE Transactions on Software Engineering, 2*(4), 308-320.

Meeker, W.Q., and Escobar, L.A. (1998). *Statistical Methods for Reliability Data.* New York: Wiley Interscience.

Meneely, A., Williams, L., Snipes, W., and Osborne, J. (2008). Predicting failures with developer networks and social network analysis. *Proceedings of the Foundations in Software Engineering* (pp. 13-23).

Menon, S., Osterman, M., and Pecht, M. (2013). *Vibration Durability of Mixed Solder Ball Grid Array Assemblies.* Prepared for the IPC Electronic Systems Technology Conference, May 21-23, Las Vegas, NV.

Miller, K.W., Morell, L.J., Noonan, R.E., Park, S.K., Nicol, D.M., Murrill, B.W., and Voas, J.M. (1992). Estimating the probability of failure when testing reveals no failures. *IEEE Transactions on Software Engineering, 18*(1), 33-43.

Mortin, D.E., Krolewski, J., and Cushing, M.J. (1995). *Consideration of Component Failure Mechanisms in the Reliability Assessment of Electronic equipment: Addressing the Constant Failure Rate Assumption.* Prepared for the Annual Reliability and Maintainability Symposium, January 16-19, Washington, DC.

Munson, J.C., and Elbaum, S. (1998). Code churn: A measure for estimating the impact of code change. *Proceedings of IEEE International Conference on Software Maintenance*, 24-31.

Musa, J. (1975). A theory of software reliability and its application. *IEEE Transactions on Software Engineering, 1*(3), 312-327.

Musa, J. (1998). *Software Reliability Engineering.* New York: McGraw-Hill.

Musa, J., and Okumoto, K. (1984). A comparison of time domains for software reliability models. *Journal of Systems and Software, 4*(4), 277-287.

Musa, J., Ianino, A., and Okumoto, K. (1987). *Software Reliability: Measurement, Prediction, Application.* New York: McGraw-Hill.

Nagappan, N., and Ball, T. (2005). Use of relative code churn measures to predict system defect density. *Proceedings of International Conference on Software Engineering*, 284-292.

Nagappan, N., Ball, T., and Murphy, B. (2006). *Using Historical In-Process and Product Metrics for Early Estimation of Software Failures.* Prepared for the International Symposium on Software Reliability Engineering, November 7-10, Raleigh, NC.

Nagappan, N., Murphy, B., and Basili, V. (2008). The influence of organizational structure on software quality: An empirical case study. *Proceedings of the International Conference on Software Engineering*, 521-530.

Nakagawa, Y., and Hanata, S. (1989). An error complexity model for software reliability measurement. *Proceedings of International Conference on Software Engineering*, 230-236.

National Research Council. (1998). *Statistics, Testing, and Defense Acquisition.* M.L. Cohen, J.E. Rolph, and D.L. Steffey, Eds. Panel on Statistical Methods for Testing and Evaluating Defense Systems, Committee on National Statistics, Commission on Behavioral and Social Sciences and Education. Washington, DC: National Academy Press.

National Research Council. (2004). *Improved Operational Testing and Evaluation and Methods of Combining Tet Information for the Stryker Family of Vehicles and Related Army Systems.* Phase II Report, Panel on Operational Test Design and Evaluation of the Interim Armored Vehicle, Committee on National Statistics. Washington, DC: The National Academies Press.

National Research Council. (2006). *Testing of Defense Systems in an Evolutionary Acquisition Environment.* Oversight Committee for the Workshop on Testing for Dynamic Acquisition of Defense Systems. V. Nair and M.L. Cohen, Eds. Committee on National Statistics, Division of Behavioral and Social Sciences and Education. Washington, DC: The National Academies Press.

Nelson, E. (1978). Estimating software reliability from test data. *Microelectronics and Reliability*, *17*(1), 67-74.

Nelson, W.B. (2003). *Recurrent Events Data Analysis for Product Repairs, Disease Recurrences, and Other Applications.* ASA-SIAM Series on Statistics and Applied Probability. Philadelphia, PA: Society for Industrial and Applied Mathematics.

Osterman, M. (2011). *Modeling Temperature Cycle Fatigue Life of SN100C Solder.* Paper prepared for the SMTA International Conference on Soldering and Reliability, May 3-4, Toronto, Canada.

Ostrand, T.J., Weyuker, E.J, and Bell, R.M. (2004). Where the bugs are. *Proceedings of the 2004 ACM SIGSOFT International Symposium on Software Testing and Analysis* (pp. 86-96).

Ostrand, T.J., E.J. Weyuker, E.J., and Bell, R.M. (2005). Predicting the location and number of faults in large software systems. *IEEE Transactions on Software Engineering, 31*(4), 340-355.

Pecht, M. (Ed.). (2009). *Product Reliability, Maintainability and Supportability Handbook.* Boca Raton, FL: CRC Press.

Pecht, M., and Dasgupta, A. (1995). Physics-of-failure: An approach to reliable product development. *Journal of the Institute of Environmental Sciences*, *38*, 30-34.

Pecht, M., and Gu, J. (2009). Physics-of-failure-based prognostics for electronic products. *Transactions of the Institute of Measurement and Control*, *31*(3/4), 309-322.

Pecht, M., and Kang, W. (1988). A critique of MIL-HDBK-217E reliability prediction methods. *IEEE Transactions on Reliability*, *37*(5), 453-457.

Pecht, M., Malhotra, A., Wolfowitz, D., Oren, M., and Cushing, M. (1992). *Transition of MIL-STD-785 from a Military to a Physics-of-Failure Based Com-Military Document.* Prepared for the 9th International Conference of the Israel Society for Quality Assurance, November 16-19, Jerusalem, Israel.

Pogdurski, A., and Clarke, L.A. (1990). A formal model of program dependences and its implications for software testing, debugging, and maintenance. *IEEE Transactions in Software Engineering, 16*(9), 965-979.

Reese, C.S., Wilson, A.G., Guo, J., Hamada, M.S., and Johnson, V. (2011). A Bayesian model for integrating multiple sources of lifetime information in system reliability assessments. *Journal of Quality Technology, 43*(2), 127-141.

Rigdon, S.E., and Basu, A P. (2000). *Statistical Methods for the Reliability of Repairable Systems*. New York: John Wiley & Sons.

Sandborn, P., Prabhakar, V., and Eriksson, B. (2008). *The Application of Product Platform Design to the Reuse of Electronic Components Subject to Long-Term Supply Chain Disruptions*. Prepared for the ASME International Design Engineering Conferences and Computers and Information in Engineering Conference, August 6, New York.

Schick, G.J., and Wolverton, R.W. (1978). An analysis of competing software reliability models. *IEEE Transactions on Software Engineering, 4*(2), 104-120.

Schröter, A., Zimmermann, T., and Zeller, A. (2006). Predicting component failures at design time. *Proceedings of International Symposium on Empirical Software Engineering*, 18-27.

Selby, R.W. (2009). Analytics-driven dashboards enable leading indicators for requirements and designs of large-scale systems. *IEEE Software, 26*(1), 41-49.

Sen, A., and Bhattacharyya, G.K. (1993). A piecewise exponential model for reliability growth and associated inferences. In A.P. Basu (Ed.), *Advances in Reliability* (pp. 331-355). Amsterdam: North-Holland.

Siegel, E. (2011). *If You Can Predict It, You Own It: Four Steps of Predictive Analytics to Own Your Market*. Prepared for the SAS Business Analytics Knowledge Exchange, June.

Siegel, E. (2012). *Uplift Modeling: Predictive Analytics Can't Optimize Marketing Decisions Without It*. White paper produced by Prediction Impact and sponsored by Pitney Bowes Business Insight, June 2011.

Sotiris, V., Tse, P.W., and Pecht, M. (2010). Anomaly detection through a Bayesian support vector machine. *IEEE Transactions on Reliability, 59*(2), 277-286.

Spencer, F.W., and Easterling, R.G. (1986). Lower confidence bounds on system reliability using component data: The Maximus methodology. In A.P. Basu (Ed.), *Reliability and Quality Control* (pp. 353-367). Amsterdam: Elsevier B.V.

Steffey, D.L., Samaniego, F.J., and Tran, H. (2000). Hierarchical Bayesian inference in related reliability experiments. In N. Limnios and M. Nikulin (Eds.), *Recent Advances in Reliability Theory: Methodology, Practice, and Inference* (pp. 379-390). Boston: Birkhauser.

Subramanyam, R., and Krishnan, M.S. (2003). Empirical analysis of CK metrics for object-oriented design complexity: Implications for software defects. *IEEE Transactions on Software Engineering, 29*(4), 297-310.

Sun, J., Cheng, S., and Pecht, M. (2012). Prognostics of multilayer ceramic capacitors via the parameter residuals. *IEEE Transactions on Device and Materials, 12*(1), 49-57.

Tahoma, Y., Tokunaga, K., Nagase, S., and Murata, Y. (1989). Structural approach to the estimation of the number of residual software faults based on the hyper-geometric distribution. *IEEE Transactions on Software Engineering, 15*(3), 345-362.

Thompson, W.E., and Chelson, P.O. (1980). On the specification and testing of software reliability. *Proceedings of the Annual Reliability and Maintainability Symposium* (pp. 379-383).

Trapnell, P.B. (1984). *Study on Reliability Fix Effectiveness Factors for Army Systems*. Technical Report 388. Aberdeen Proving Ground, MD: U.S. Army Materiel Systems Analysis Activity.

U.S. Department of Defense. (1982). *Test and Evaluation of System Reliability, Availability, and Maintainability: A Primer*. Report No. DoD 3235.1-H. Washington, DC: Author.

U.S. Department of Defense. (1991). *MIL-HDBK-217. Military Handbook: Reliability Prediction of Electronic Equipment.* Washington, DC: U.S. Department of Defense. Available: http://www.sre.org/pubs/Mil-Hdbk-217F.pdf [August 2014].

U.S. Department of Defense. (2004). *Department of Defense Directive 4151.18.* Subject: Maintenance of Military Materiel, March 31.

U.S. Department of Defense. (2005). *DoD Guide for Achieving Reliability, Availability, and Maintainability.* Washington, DC: Author.

U.S. Department of Defense. (2008a). *Report of the Defense Science Board Task Force on Developmental Test and Evaluation.* Washington DC: Office of the Secretary of Defense for Acquisition, Technology, and Logistics.

U.S. Department of Defense. (2008b). *Operation of the Defense Acquisition System.* DoD Instruction Number 5000.02. Washington, DC: Author.

U.S. Department of Defense. (2008c). *Report of the Reliability Improvement Working Group.* Washington, DC: Author.

U.S. Department of Defense. (2010). *Guidance on the Use of Design of Experiments (DOE) in Operational Test and Evaluation.* Director, Operational Test and Evaluation. Available: http://www.dote.osd.mil/pub/reports/20101019GuidanceonuseofDOEinOT&E.pdf [October 2014].

U.S. Department of Defense. (2011a). *Department of Defense Handbook: Reliability Growth Management.* MIL-HDBK-189C. Washington, DC: Author.

U.S. Department of Defense. (2011b). *FY 2011 Annual Report.* Director, Operational Test and Evaluation. Available: http://www.dote.osd.mil/pub/reports/FY2012 [October 2014].

U.S. Department of Defense. (2012). *Test and Evaluation Management Guide.* Washington, DC: Author.

U.S. Department of Defense. (2013a). *Defense Acquisition Guidebook.* Washington, DC: Author.

U.S. Department of Defense. (2013b). *Directive-Type Memorandum (DTM 11-003)–Reliability Analysis, Planning, Tracking, and Reporting.* Washington, DC: Author.

U.S. Government Accountability Office. (2008). *Best Practices: Increased Focus on Requirements and Oversight Needed to Improve DOD's Acquisition Environment and Weapon System Quality.* GAO-08-294. Washington, DC: Author.

Vasan, A., Long, B., and Pecht, M. (2012). Diagnostics and prognostics method for analog electronic circuits. *IEEE Transactions on Industrial Electronics, 60*(11), 5277-5291.

Vouk, M.A., and Tai, K.C. (1993). Multi-phase coverage- and risk-based software reliability modeling. *Proceedings of CASCON '93*, 513-523.

Walls, L., Quigley, J., and Krasich, M. (2005). Comparison of two models for managing reliability growth during product design. *IMA Journal of Management Mathematics, 16*(1), 12-22.

Weiss, S.N., and Weyuker, E.J. (1988). An extended domain-bases model of software reliability. *IEEE Transactions on Software Engineering, 14*(10), 1512-1524.

Weyuker, E.J., and Ostrand, T., and Bell, R. (2008). Do too many cooks spoil the broth? Using the number of developers to enhance defect prediction models. *Empirical Software Engineering Journal*, October.

Whittaker, J.A. (1992). *Markov Chain Techniques for Software Testing and Reliability Analysis* (Ph.D. dissertation). University of Tennessee, Department of Computer Science, Knoxville.

Whittaker, J.A., and Poore, J.H. (1993). Markov analysis of software specifications. *ACM Transactions on Software Engineering and Methodology, 2*(1), 93-106.

Wilson, A.G., Graves, T., Hamada, M.S., and Reese, C.S. (2006). Advances in data combination and collection for system reliability assessment. *Statistical Science, 21*(4), 514-531.

Wohlin, C., and Korner, U. (1990). Software faults: Spreading, detection and costs. *Software Engineering Journal*, *5*(1), 38-42.

Wong, K.L. (1990). What is wrong with the existing reliability prediction methods? *Quality and Reliability Engineering International*, *6*(4), 251-258.

Xie, M. (1991). *Software Reliability Modeling*. Singapore: World Scientific.

Yamada, S., and Osaki, S. (1983). Software reliability growth modeling: Models and applications. *IEEE Transactions on Software Engineering*, *11*(12), 1431-1437.

Yamada, S., and Osaki, S. (1985). Discrete software reliability growth models. *Journal of Applied Stochastic Models and Data Analysis*, *1*(4).

Zimmermann, T., and Nagappan, N. (2008). Predicting defects using network analysis on dependency graphs. *Proceedings of the International Conference on Software Engineering* (pp. 531-540).

Zimmermann, T., Weissgerber, P., Diehl, S., and Zeller, A. (2005). Mining version histories to guide software changes. *IEEE Transactions in Software Engineering*, *31*(6), 429-445.

Appendix A

Recommendations of Previous Relevant Reports of the Committee on National Statistics

The Committee on National Statistics has carried out a number of studies sponsored by the Under Secretary of Defense for Acquisition, Technology, and Logistics and the Director of Operational Test and Evaluation of the U.S. Department of Defense that are relevant to this study. The previous studies covered the application of statistical, system engineering, and software engineering techniques to improve the development of defense systems. Many of the conclusions and recommendations of these studies are relevant to the development of reliable defense systems, the topic of this study.

The rest of this appendix reproduces those conclusions and recommendations, a large number of which have not been fully implemented. Their inclusion here serves both to highlight that some of the issues in this report have a long history and to emphasize the connections among the many parts of the system of defense development, acquisition, and testing.

The reports are listed in chronological order; the conclusions and recommendations are produced in full. All the reports were published by and are available from the National Academies Press.

Statistics, Testing, and Defense Acquisition: New Approaches and Methodological Improvements (1998)

This study was a general overview of the application of statistical methods to many components of defense acquisition.

RECOMMENDATION 7.1 The Department of Defense and the military services should give increased attention to their reliability, availability, and maintainability data collection and analysis procedures because deficiencies continue to be responsible for many of the current field problems and concerns about military readiness. (p. 105)

RECOMMENDATION 7.2 The Test and Evaluation Master Plan and associated documents should include more explicit discussion of how reliability, availability, and maintainability issues will be addressed, particularly the extent to which operational testing, modeling, simulation, and expert judgment will be used. The military services should make greater use of statistically designed tests to assess reliability, availability, and maintainability and related measures of operational suitability. (p. 106)

RECOMMENDATION 7.3 As part of an increased emphasis on assessing the reliability, availability, and maintainability of prospective defense systems, reasonable criteria should be developed for each system. Such criteria should permit a balancing of a variety of considerations and be explicitly linked to estimates of system cost and performance. A discussion of the implications for performance, and cost of failing, if the system's demonstrated reliability, availability, and maintainability characteristics fall below numerical goals should be included. (p. 107)

RECOMMENDATION 7.4 Operational test agencies should promote more critical attention to the specification of statistical models of equipment reliability, availability, and maintainability and to supporting the underlying assumptions. Evidence from plots, diagnostics, and formal statistical tests—developed from the best currently available methods and software—should be used to justify the choice of statistical models used in both the design and the analysis of operational suitability tests. (p. 113)

RECOMMENDATION 7.6 Service test agencies should carefully document, in advance of operational testing, the failure definitions and criteria to be used in scoring reliability, availability, and maintainability data. The objectivity of the scoring procedures that were actually implemented should be assessed and included in the reporting of results. The sensitivity of final reliability, availability, and maintainability estimates to plausible alternative interpretations of test data, as well as subsequent assumptions concerning operating tempo and logistics support, should be discussed in the reporting. (p. 116)

RECOMMENDATION 7.7 Methods of combining reliability, availability, and maintainability data from disparate sources should be carefully studied and selectively adopted in the testing processes associated with the Department of Defense acquisition programs. In particular, authorization should be given to operational testers to combine reliability, availability, and maintainability data from developmental and operational testing as appropriate, with the proviso that analyses in which this is done be carefully justified and defended in detail. (p. 119)

RECOMMENDATION 7.8 All service-approved reliability, availability, and maintainability data, including vendor-generated data, from technical, developmental, and operational tests, should be properly archived and used in the final preproduction assessment of a prospective system. After procurement, field performance data and associated records should be retained for the system's life, and used to provide continuing assessment of its reliability, availability, and maintainability characteristics. (p. 120)

RECOMMENDATION 7.9 Any use of model-based reliability predictions in the assessment of operational suitability should be validated a posteriori with test and field experience. Persistent failure to achieve validation should contraindicate the use of reliability growth models for such purposes. (p. 122)

RECOMMENDATION 7.10 Given the potential benefits of accelerated reliability testing methods, we support their further examination and use. To avoid misapplication, any model that is used as an explicit or implicit component of an accelerated reliability test must be subject to the same standards of validation, verification, and certification as models that are used for evaluation of system effectiveness. (p. 124)

RECOMMENDATION 7.11 The Department of Defense should move aggressively to adapt for all test agencies the Organization for International Standardization (ISO) standards relating to reliability, availability, and maintainability. Great attention should be given to having all test agencies ISO-certified in their respective areas of responsibility for assuring the suitability of prospective military systems. (p. 125)

RECOMMENDATION 7.12 Military reliability, availability, and maintainability testing should be informed and guided by a new battery of military handbooks containing a modern treatment of all pertinent topics in the fields of reliability and life testing, including, but not limited to, the design and analysis of standard and accelerated tests, the

handling of censored data, stress testing, and the modeling of and testing for reliability growth. The modeling perspective of these handbooks should be broad and include practical advice on model selection and model validation. The treatment should include discussion of a broad array of parametric models and should also describe nonparametric approaches. (p. 126)

Innovations in Software Engineering for Defense Systems (2003)

This study covered statistically oriented software engineering methods that were useful for the development of defense systems.

RECOMMENDATION 1 **Given the current lack of implementation of state-of-the-art methods in software engineering in the service test agencies, initial steps should be taken to develop access to—either in-house or in a closely affiliated relationship—state-of-the-art software engineering expertise in the operational or developmental service test service agencies.** (p. 2)

RECOMMENDATION 2 **Each service's operational or developmental test agency should routinely collect and archive data on software performance, including test performance data and data on field performance. The data should include fault types, fault times and frequencies, turnaround rate, use scenarios, and root cause analysis. Also, software acquisition contracts should include requirements to collect such data.** (p. 3)

RECOMMENDATION 3 **Each service's operational or developmental test agency should evaluate the advantages of the use of state-of-the-art procedures to check the specification of requirements for a relatively complex defense software-intensive system.** (p. 3)

RECOMMENDATION 6 **DoD needs to examine the advantages and disadvantages of the use of methods for obligating software developers under contract to DoD to use state-of-the-art methods for requirements analysis and software testing, in particular, and software engineering and development more generally.** (p. 4)

Testing of Defense Systems in an Evolutionary Acquisition Environment (2006)

This study considered the methods that would be applicable to defense systems development that is implemented in stages.

1. . . . revise DoD testing procedures to explicitly require that developmental tests have an operational perspective . . . ,
2. require (DoD) contractors to share all relevant data on system performance and the results of modeling and simulation . . .

Industrial Methods for the Effective Development and Testing of Defense Systems (2012)

This study examined the potential of the use of system and software engineering methods for defense system development.

> The performance of a defense system early in development is often not rigorously assessed, and in some cases the results of assessments are ignored; this is especially so for suitability assessments. This lack of rigorous assessment occurs in the generation of system requirements; in the timing of the delivery of prototype components, subsystems, and systems from the developer to the government for developmental testing; and in the delivery of production-representative system prototypes for operational testing. As a result, throughout early development, systems are allowed to advance to later stages of development when substantial design problems remain.

Appendix B

Workshop Agenda

WORKSHOP ON RELIABILITY GROWTH

Panel on the Theory and Application of
Reliability Growth Modeling to Defense Systems
Committee on National Statistics
National Academy of Sciences

September 22-23, 2011

Holiday Inn Washington Central
Mayor's Room
1501 Rhode Island Avenue NW
Washington, DC 20005

AGENDA

Thursday, September 22

8:45 am Key Issues in Reliability Growth
Frank Kendall, Acquisitions, Technology, and Logistics, DoD; and Michael Gilmore, Office of the Director, Operational Testing and Evaluation, DoD

9:30 am	Implementation Strategy upon Approval of DoD Reliability Policy DTM 11-003 Andy Monje, Mission Assurance, Office of the Deputy Assistant Secretary of Defense for Systems Engineering, DoD
10:00 am	Review of Reliability Management, Design and Growth Standards Available to DoD and Industry—Reliability Information Analysis Center David Nicholls, Reliability Information Analysis Center
10:25 am	Further Views on GEIA-STD 0009 Paul Shedlock, Raytheon
10:50 am	Break
11:00 am	A View from Defense Contractors Tom Wissink, Lockheed Martin, and Lou Gullo, Raytheon
12:10 pm	Lunch
1:10 pm	View from Nondefense Contractors Guangbin Yang, Ford; Shirish Kher, Alcatel-Lucent; and Martha Gardner, General Electric
2:20 pm	ATEC [U.S. Army Test and Evaluation Command] Reliability Growth Case Studies and Lessons Learned Mike Cushing, Army Evaluation Center
3:15 pm	Defense Experiences Albert (Bud) Boulter, Air Force Operational Test and Evaluation Center; James Woodford, Chief Systems Engineer, Research, Development, and Acquisitions, U.S. Navy; and Karen Bain, U.S. Navy Air Systems Command
4:30 pm	Break
4:45 pm	Some Software Complications in Reliability Assessment William McCarthy, Operational Test and Evaluation Force, U.S. Navy, and Patrick Sul, Office of the Director, Operational Test and Evaluation, DoD

5:15 pm Software Testing
Nachi Nagappan, Microsoft

5:40 pm Floor Discussion

6:00 pm Adjourn

Friday, September 23

8:45 am Testing to Assess Reliability and Other Design Issues for Hardware Systems
E.A. Elsayed, Department of Industrial and Systems Engineering, Rutgers University

9:25 am The DT/OT Gap
Paul Ellner, Army Test and Evaluation Command

10:00 am Break

10:10 am Reconsidering the Foundations of Reliability Theory
Nozer Singpurwalla, Department of Statistics, George Washington University

10:40 am Reliability Growth and Beyond
Don Gaver, Operations Research Department, Naval Postgraduate School

11:10 am A New Look at Fix Effectiveness Factors
Steve Brown, Commercial Engineering, Lennox International

11:40 am Various Technical Issues: Open Discussion
Moderator: Ananda Sen, Department of Biostatistics, University of Michigan

12:00 pm Working Lunch (with continuing discussion)

1:00 pm Adjourn

Appendix C

Recent DoD Efforts to Enhance System Reliability in Development

Since the early 2000s, the U.S. Department of Defense (DoD) has carried out several examinations of the defense acquisition process and has begun to develop some new approaches. These developments were in response to what was widely viewed as deterioration in the process over the preceding two or so decades.

During the 1990s, some observers characterized the view of defense acquisition as one of believing that if less oversight were exercised over the development of weapons systems by contractors, the result would be higher quality (and more timely delivery of) defense systems. Whether or not this view was widely held, the 1990s were also the time that a large fraction of the reliability engineering expertise in both the Office of the Secretary of Defense (OSD) and the services was lost (see Adolph et al., 2008), and some of the formal documents providing details on the oversight of system suitability were either cancelled or not updated. For instance, DoD's Guide *for Achieving Reliability, Availability, and Maintainability* (known as the RAM Primer) failed to include many newer methods, the *Military Handbook: Reliability Prediction of Electronic Equipment* (MIL-HDBK-217) became increasingly outdated, and in 1998 DoD cancelled *Reliability Program for Systems and Equipment Development and Production* (MIL-STD-785B) (although industry has continued to follow its suggested procedures).[1]

[1]A criticism of this standard was that it took too reactive an approach to achieving system reliability goals. In particular, this standard presumed that approximately 30 percent of system reliability would come from design choices, while the remaining 70 percent would be achieved through reliability growth during testing

During the early to mid-2000s, it became increasingly clear that something, possibly this strategy of relaxed oversight, was not working, at least insofar as the reliability of fielded defense systems. Summaries of the evaluations of defense systems in development in the annual reports of the Director of Operational Testing and Evaluation (DOT&E) between 2006 and 2011 reported that a large percentage of defense systems—often as high as 50 percent—failed to achieve their required reliability during operational test. Because of the 10- to 15-year development time for defense systems, relatively recent data still reflect the procedures in place during the 1990s.

The rest of this appendix highlights sections from reports over the past 8 years that have addressed this issue.

DOT&E 2007 ANNUAL REPORT

In the 2007 DOT&E Annual Report, Director Charles McQueary provided a useful outline of what, in general, needed to be changed to improve the reliability of defense systems, and in particular why attention to reliability early in development was important (U.S. Department of Defense, 2007a, p. i):

> Contributors to reliability problems include: poor definition of reliability requirements, a lack of understanding by the developer on how the user will operate and maintain the system when fielded, lack of reliability incentives in contracting, and poor tracking of reliability growth during system development.

He also wrote that addressing such concerns was demonstrably cost-beneficial and that best practices should be identified (pp. i-ii):

> [O]ur analysis revealed reliability returns-on-investment between a low of 2 to 1 and a high of 128 to 1. The average expected return is 15 to 1, implying a $15 savings in life cycle costs for each dollar invested in reliability. . . . Since the programs we examined were mature, I believe that earlier reliability investment (ideally, early in the design process), could yield even larger returns with benefits to both warfighters and taxpayers. . . . I also believe an effort to define best practices for reliability programs is vital and that these should play a larger role in both the guidance for, and the evaluation of, program proposals. Once agreed upon and codified, reliability program standards could logically appear in both Requests for Proposals (RFPs) and, as appropriate, in contracts. Industry's role is key in this area.

2008 REPORT OF THE DEFENSE SCIENCE BOARD

In May 2008, the DoD's Defense Science Board's Task Force on Developmental Test and Evaluation issued its report (U.S. Department of Defense, 2008a), which included the following assertions about defense system development (p. 6):

> The single most important step necessary to correct high suitability failure rates is to ensure programs are formulated to execute a viable systems engineering strategy from the beginning, including a robust reliability, availability, and maintainability (RAM) program, as an integral part of design and development. No amount of testing will compensate for deficiencies in RAM program formulation.

The report found that the use of reliability growth in development had been discontinued by DoD more than 15 years previously. It made several recommendations to the department regarding the defense acquisition process (p. 6):

> [DoD should] identify and define RAM requirements within the Joint Capabilities Integration Development System (JCIDS), and incorporate them in the Request for Proposal (RFP) as a mandatory contractual requirement . . . during source selection, evaluate the bidder's approaches to satisfying RAM requirements. Ensure flow-down of RAM requirements to subcontractors, and require development of leading indicators to ensure RAM requirements are met.

In addition, the task force recommended that DoD require (p. 6)

> [the inclusion of] a robust reliability growth program, a mandatory contractual requirement and document progress as part of every major program review . . . [and] ensure that a credible reliability assessment is conducted during the various stages of the technical review process and that reliability criteria are achievable in an operational environment.

This report also argued that there was a need for a standard of best practices that defense contractors could use to prepare proposals and contracts for the development of new systems. One result of this suggestion was the formation of a committee that included representatives from industry, DoD, academia, and the services, under the auspices of the Government Electronics and Information Technology Association (GEIA). The resulting standard, ANSI/GEIA-STD-0009, "Reliability Program Standard for Systems Design, Development, and Manufacturing,"[2] was certified by the

[2]This standard replaced MIL-STD-785, Reliability Program for Systems and Equipment.

American National Standards Institute in 2008 and designated as a DoD standard to make it easy for program managers to incorporate best reliability practices in requests for proposals (RFPs) and in contracts.

ARMY ACQUISITION EXECUTIVE MEMORANDUM

About the same time, the Army modified its acquisition policy, described in an Army acquisition executive memorandum on reliability (U.S. Department of Defense, 2007b). The memo stated (p. 1): "Emerging data shows that a significant number of U.S. Army systems are failing to demonstrate established reliability requirements during operational testing and many of these are falling well short of their established requirement."

To address this problem, the Army instituted a system development and demonstration reliability test threshold process. The process mandated that an initial reliability threshold be established early enough to be incorporated into the system development and demonstration contract. It also said that the threshold should be attained by the end of the first full-up, integrated, system-level developmental test event. The default value for the threshold was 70 percent of the reliability requirement specified in the capabilities development document. Furthermore, the Test and Evaluation Master Plan (TEMP)[3] was to include test and evaluation planning for evaluation of the threshold and growth of reliability throughout system development.[4] Also about this time, the Joint Chiefs of Staff published an updated instruction about system requirements (CJCSI 3170.01F)[5] that declared materiel availability, a component of suitability that is a function of reliability, a "key performance parameter."

RELIABILITY IMPROVEMENT WORKING GROUP

Also in 2008, DOT&E established a Reliability Improvement Working Group, with three goals: ensuring that each DoD acquisition program incorporates a viable systems engineering strategy, including a RAM growth program; promoting the reconstitution of cadres of experienced test and evaluation and RAM personnel across government organizations; and implementing mandated integrated developmental and operational testing, including the sharing of and access to all appropriate contractor and

---

[3]The TEMP is the high-level "basic planning document for all life cycle Test and Evaluation (T&E) that are related to a particular system acquisition and is used by all decision bodies in planning, reviewing, and approving T&E activity" (U.S. Department of the Army, Pamphlet 73-2, 1996, p. 1).

[4]Many of these initiatives are described in the succession of DOT&E Annual Reports.

[5]Available: http://www.dtic.mil/cjcs_directives/cdata/unlimit/3170_01.pdf [August 2014].

government data and the use of operationally representative environments in early testing.

The subsequent report of this working group (U.S. Department of Defense, 2008b) argued for six requirements: (1) a mandatory reliability policy, (2) program guidance for early reliability planning, (3) language for RFPs and contracts, (4) a scorecard to evaluate bidders' proposals, (5) standard evaluation criteria for credible assessments of program progress, and (6) the hiring of a cadre of experts in each service. The report also endorsed use of specific contractual language that was based on that in ANSI/GEIA-STD-0009 (see above).

With respect to RFPs, the working group report contained the following advice for mandating reliability activities in acquisition contracts (U.S. Department of Defense, 2008b, p. II-6):

> The contractor shall develop a reliability model for the system. At minimum, the system reliability model shall be used to (1) generate and update the reliability allocations from the system level down to lower indenture levels, (2) aggregate system-level reliability based on reliability estimates from lower indenture levels, (3) identify single points of failure, and (4) identify reliability-critical items and areas where additional design or testing activities are required in order to achieve the reliability requirements. The system reliability model shall be updated whenever new failure modes are identified, failure definitions are updated, operational and environmental load estimates are revised, or design and manufacturing changes occur throughout the life cycle. Detailed component stress and damage models shall be incorporated as appropriate.

The report continued with detailed requirements for contractors (pp. II-6, 7):

> The contractor shall implement a sound systems-engineering process to translate customer/user needs and requirements into suitable systems/products while balancing performance, risk, cost, and schedule. . . . The contractor shall estimate and periodically update the operational and environmental loads (e.g., mechanical shock, vibration, and temperature cycling) that the system is expected to encounter in actual usage throughout the life cycle. These loads shall be estimated for the entire life cycle which will typically include operation, storage, shipping, handling, and maintenance. The estimates shall be verified to be operationally realistic with measurements using the production-representative system in time to be used for Reliability Verification. . . . The contractor shall estimate the lifecycle loads that subordinate assemblies, subassemblies, components, commercial-off-the-shelf, non-developmental items, and government-furnished equipment will experience as a result of the product-level operational and environmental loads estimated above. These estimates and updates shall be provided to teams developing assemblies, subassemblies,

and components for this system. . . . The identification of failure modes and mechanisms shall start immediately after contract award. The estimates of lifecycle loads on assemblies, subassemblies, and components obtained above shall be used as inputs to engineering- and physics-based models in order to identify potential failure mechanisms and the resulting failure modes. The teams developing assemblies, subassemblies, and components for this system shall identify and confirm through analysis, test or accelerated test the failure modes and distributions that will result when lifecycle loads estimated above are imposed on these assemblies, subassemblies and components. . . . All failures that occur in either test or in field shall be analyzed until the root cause failure mechanism has been identified. Identification of the failure mechanism provides the insight essential to the identification of corrective actions, including reliability improvements. Predicted failure modes/mechanisms shall be compared with those from test and the field. . . . The contractor shall have an integrated team, including suppliers of assemblies, subassemblies, components, commercial-off-the-shelf, non-developmental items, and government-furnished equipment, as applicable, analyze all failure modes arising from modeling, analysis, test, or the field throughout the life cycle in order to formulate corrective actions. . . . The contractor shall deploy a mechanism (e.g., a Failure Reporting, Analysis, and Corrective Action System or a Data Collection, Analysis, and Corrective Action System) for monitoring and communicating throughout the organization (1) description of test and field failures, (2) analyses of failure mode and root-cause failure mechanism, (3) the status of design and/or process corrective actions and risk-mitigation decisions, (4) the effectiveness of corrective actions, and (5) lessons learned. . . . The model developed in System Reliability Model shall be used, in conjunction with expert judgment, in order to assess if the design (including commercial-off-the-shelf, non-developmental items, and government-furnished equipment) is capable of meeting reliability requirements in the user environment. If the assessment is that the customer's requirements are infeasible, the contractor shall communicate this to the customer. The contractor shall allocate the reliability requirements down to lower indenture levels and flow them and needed inputs down to its subcontractors/suppliers. The contractor shall assess the reliability of the system periodically throughout the life cycle using the System Reliability Model, the lifecycle operational and environmental load estimates generated herein, and the failure definition and scoring criteria. . . . The contractor shall understand the failure definition and scoring criteria and shall develop the system to meet reliability requirements when these failure definitions are used and the system is operated and maintained by the user. . . . The contractor shall conduct technical interchanges with the customer/user in order to compare the status and outcomes of Reliability Activities, especially the identification, analysis, classification, and mitigation of failure modes.

REQUIREMENTS FOR TEMPS

Also beginning in 2008, DOT&E initiated the requirement that TEMPs contain a process for the collection and reporting of reliability data and that they present specific plans for reliability growth during system development. The 2011 DOT&E Annual Report (U.S. Department of Defense, 2011a) reported on the effect of this requirement, noting that in a survey of 151 programs with DOT&E-approved TEMPs in development carried out in 2010 and focusing on those programs with TEMPs approved since 2008, 90 percent planned to collect and report reliability data. (There have been more recent reviews carried out by DOT&E with similar results; see, in particular U.S. Department of Defense, 2013.) In addition, these TEMPs were more likely to: (1) have an approved system engineering plan, (2) incorporate reliability as an element of test strategy, (3) document their reliability growth strategy in the TEMP, (4) include reliability growth curves in the TEMP, (5) establish reliability-based milestone or operational testing entrance criteria, and (6) collect and report reliability data. Unfortunately, possibly because of the long development time for defense systems, or possibly because of a disconnect between reporting and practice, there has as yet been no significant improvement in the percentage of such systems that meet their reliability thresholds. Also, there is no evidence that programs are using reliability metrics to ensure that the growth in reliability will result in the system's meeting their required levels. As a result, systems continue to enter operational testing without demonstrating their required reliability.

LIFE-CYCLE COSTS AND RAM REQUIREMENTS

The Defense Science Board (U.S. Department of Defense, 2008a) study on developmental test and evaluation also helped to initiate four activities: (1) establishment of the Systems Engineering Forum, (2) institution of reliability growth training, (3) establishment of a reliability senior steering group, and (4) establishment of the position of Deputy Assistant Secretary of Defense (System Engineering).

The Defense Science Board's study also led to a memorandum from the Under Secretary of Defense for Acquisition, Technology, and Logistics (USD AT&L) "Implementing a Life Cycle Management Framework" (U.S. Department of Defense, 2008c). This memorandum directed the service secretaries to establish policies in four areas to carry out the following.

First, all major defense acquisition programs were to establish target goals for the metrics of materiel reliability and ownership costs. This was to be done through effective collaboration between the requirements and acquisition communities that balanced funding and schedule while ensuring system suitability in the anticipated operating environment. Also, resources were to be aligned to achieve readiness levels. Second, reliability performance was to be tracked throughout the program life cycle. Third, the services were to ensure that development contracts and acquisition plans evaluated RAM during system design. Fourth, the Services were to evaluate the appropriate use of contract incentives to achieve RAM objectives.

The 2008 DOT&E Annual Report (U.S. Department of Defense, 2008d, p. iii) stressed the new approach:

> . . . a fundamental precept of the new T&E [test and evaluation] policies is that expertise must be brought to bear at the beginning of the system life cycle to provide earlier learning. Operational perspective and operational stresses can help find failure modes early in development when correction is easiest. A key to accomplish this is to make progress toward Integrated T&E, where the operational perspective is incorporated into all activity as early as possible. This is now policy, but one of the challenges remaining is to convert that policy into meaningful practical application.

In December 2008, the USD AT&L issued an "Instruction" on defense system acquisition (U.S. Department of Defense, 2008e). It modified DODI 5000.02[6] by requiring that program managers should formulate a "viable RAM strategy that includes a reliability growth program as an integral part of design and development" (p. 19). It stated that RAM was to be integrated within the Systems Engineering processes, as documented in the program's Systems Engineering Plan (SEP) and Life-Cycle Sustainment Plan (LCSP), and progress was to be assessed during technical reviews, test and evaluation, and program support reviews. It stated that (p. vi)

> For this policy guidance to be effective, the Services must incorporate formal requirements for early RAM planning into their regulations, and assure development programs for individual systems include reliability growth and reliability testing; ultimately, the systems have to prove themselves in operational testing. Incorporation of RAM planning into Service regulation has been uneven.

In 2009, the Weapons System Acquisition Reform Act (WSARA, P.L. 111-23) required that acquisition programs develop a reliability growth program.[7] It prescribed that the duties of the directors of systems engineering were to develop policies and guidance for "the use of systems

[6]See discussion in Chapter 9. Instruction 5000.02, Operation of the Defense Acquisition System, is available at http://www.dtic.mil/whs/directives/corres/pdf/500002_interim.pdf [December 2013].

[7]P.L. 111-23 is available at http://www.acq.osd.mil/se/docs/PUBLIC-LAW-111-23-22MAY2009.pdf [January 2014].

engineering approaches to enhance reliability, availability, and maintainability on major defense acquisition programs, [and] the inclusion of provisions relating to systems engineering and reliability growth in requests for proposals" (section 102).

WSARA (U.S. Department of Defense, 2009a) also stated that adequate resources needed to be provided and should (section 102):

> . . . include a robust program for improving reliability, availability, maintainability, and sustainability as an integral part of design and development, . . . [and] identify systems engineering requirements, including reliability, availability, maintainability, and lifecycle management and sustainability requirements, during the Joint Capabilities Integration Development System (JCIDS) process, and incorporate such systems engineering requirements into contract requirements for each major defense acquisition program.

Shortly after WSARA was adopted, a RAM cost (RAM-C) manual was produced (U.S. Department of Defense, 2009b) to guide the development of realistic reliability, availability, and maintainability requirements for the established suitability/sustainability key performance parameters and key system attributes. The RAM-C manual contains

- RAM planning and evaluation tools, first to assess the adequacy of the RAM program proposed and then to monitor the progress in achieving program objectives. In addition, DoD has sponsored the development of tools to estimate the investment in reliability that is needed and the return on investment possible in terms of the reduction of total life-cycle costs. These tools include algorithms to estimate how much to spend on reliability.
- Workforce and expertise initiatives to bring back personnel with the expertise that was lost during the years that the importance of government oversight of RAM was discounted.

In 2010, DOT&E published sample RFP and contract language to help assure reliability growth was incorporated in system design and development contracts. DOT&E also sponsored the development of the reliability investment model (see Forbes et al., 2008) and began drafting the Reliability Program Handbook, HB-0009, meant to assist with the implementation of ANSI/GEIA-STD-0009. The TechAmerica Engineering Bulletin Reliability Program Handbook, TA-HB-0009 was released and published in May 2013.

On June 30, 2010, DOT&E issued a memorandum, "State of Reliability," which strongly argued that sustainment costs are often much more than 50 percent of total system costs and that unreliable systems have much higher sustainment costs because of the need for spare systems, increased maintenance, increased number of repair parts, more repair facilities, and more staff (U.S. Department of Defense, 2010). Also, poor reliability hinders warfighter effectiveness (pp. 1-2):

> For example, the Early-Infantry Brigade Combat Team (E-IBCT) unmanned aerial system (UAS) demonstrated a mean time between system aborts of 1.5 hours, which was less than 1/10th the requirement. It would require 129 spare UAS to provide sufficient number to support the brigade's operations, which is clearly infeasible. When such a failing is discovered in post-design testing—as is typical with current policy—the program must shift to a new schedule and budget to enable redesign and new development. For example, it cost $700M to bring the F-22 reliability up to acceptable levels.

However, this memo also points out that increases in system reliability can also come at a cost. A reliable system can weigh more, and be more expensive, and sometimes the added reliability does not increase battlefield effectiveness and is therefore wasteful.

The memo also discussed the role of contractors (p. 2):

> [Industry] will not bid to deliver reliable products unless they are assured that the government expects and requires all bidders to take the actions and make the investments up-front needed to develop reliable systems. To obtain reliable products, we must assure vendors' bids to produce reliable products outcompete the cheaper bids that do not.

The memo also stressed that reliability constraints must be "pushed as far to the left as possible," meaning that the earlier that design-related reliability problems are discovered, the less expensive it is to correct such problems, and the less impact there is on the completion of the system. Finally, the memo stated that all DoD acquisition contracts will require, at a minimum, the system engineering practices of ANSI/GEIA STD-0009 (Information Technology Association of America, 2008).

TWO MAJOR INITIATIVES TO PROMOTE RELIABILITY GROWTH: ANSI/GEIA-STD-0009 AND DTM 11-003

ANSI/GEIA-STD-0009: A Standard to Address Reliability Deficiencies

ANSI/GEIA-STD-0009 (Information Technology Association of America, 2008) is a recent document that can be agreed to be used as a standard for DoD purposes. It begins with a statement that the user's needs are represented by four reliability objectives (p. 1):

1. The developer shall solicit, investigate, analyze, understand, and agree to the user's requirements and product needs.
2. The developer shall use well-defined reliability- and systems-engineering processes to develop, design, and verify that the system/product meets the user's documented reliability requirements and needs.
3. The multifunctional team shall verify during production that the developer has met the user's reliability requirements and needs prior to fielding.
4. The multifunctional team shall monitor and assess the reliability of the system/product in the field.

ANSI/GEIA-STD-0009 provides detailed advice on what information should be mandated for inclusion in several documents provided by the contractor, including a reliability program plan (RPP), in order to satisfy the above four objectives. To satisfy Objective 1 to understand customer/user requirements and constraints, the RPP shall (Information Technology Association of America, 2008, p. 15):

- Define all resources (e.g., personnel, funding, tools, and facilities) required to fully implement the reliability program.
- Include a coordinated schedule for conducting all reliability activities throughout the system/product life cycle.
- Include detailed descriptions of all reliability activities, functions, documentation, processes, and strategies required to ensure system/product reliability maturation and management throughout the system/product life cycle.
- Document the procedures for verifying that planned activities are implemented and for both reviewing and comparing their status and outcomes.
- Manage potential reliability risks due, for example, to new technologies or testing approaches.
- Ensure that reliability allocations, monitoring provisions, and inputs that impact reliability (e.g., user and environmental loads) flow down to subcontractors and suppliers.
- Include contingency-planning criteria and decision making for altering plans and intensifying reliability improvement efforts.
- Include, at minimum, the normative activities identified throughout this standard.
- Include, when applicable, additional customer-specified normative activities.

Furthermore, the standard says that the RPP "shall address the implementation of all the normative activities identified in Objectives 1-4." This standard requires that the RFP call for the inclusion in acquisition proposals of a description of system or product reliability model and requirements, which means the description of the methods and tools used, the extent to which detailed stress and damages models will be employed, how and when models and requirements will be updated in response to design evolution and the discovery of failure modes, and how the model and requirements will be used to prioritize design elements. This standard also requires that proposals include a description of the engineering process, which includes how reliability improvements will be incorporated in the design, how it will be ensured that design rules that impact reliability will be adhered to, how reliability-critical items will be identified, managed, and controlled, and how the reliability impact of design changes will be monitored and evaluated. In addition, the standard calls for proposals to include the assessment of life-cycle loads, the impact of those loads on subsystems and components, the identification of failure modes and mechanisms, the description of a closed-loop failure-mode mitigation process, how and when reliability assessments will be performed, plan design, production, and field reliability verification, failure definitions and scoring, technical reviews, and outputs and documentation.

With reference to the last point, we note that life-cycle loads may be difficult to predict. For example, a truck that is designed to be reliable in cross-country maneuvers may be less reliable on sustained highway travel. In general, a system's actual life cycle may include new missions for the system to carry out. For some systems, it might be appropriate to conclude from testing that it is reliable for some scenarios and not others. Such a statement would be similar to statements that the system is effective in certain operational situations but not others. It may be that system reliability for all possible missions is too expensive; perhaps there should be different reliability requirements for different missions.

This standard provides greater detail on the satisfaction of Objective 2: design and redesign for reliability. The goal is to ensure the use of well-defined reliability engineering processes to develop, design, manufacture, and sustain the system/product so that it meets the user's reliability requirements and needs. This includes the initial conceptual reliability model of the system, quantitative reliability requirements for the system, initial reliability assessment, user and environmental life-cycle loads, failure definitions and scoring criteria, the reliability program plan, and the reliability requirements verification strategy. Furthermore, this also includes updates to the RPP, refinements to the reliability model, including reliability allocations to subsystems and components, refined user and environmental loads, initial estimates of loads for subsystems and components, engineering analysis and test data identifying the system failure modes that will result from life-cycle loads, data verifying the mitigation of these failure modes, updates of the reliability requirements verification strategy, and updates to the reliability assessment.

The standard also says that the developer should develop a model that relates component-level reliabilities to system-level reliabilities. In addition, the identification of failure modes and mechanisms shall start as soon as the development begins. Failures that occur in either test or the field are to be analyzed until the root-cause failure mechanism has been identified. In addition, the developer must make use of a closed-loop failure mitigation process. The developer (p. 26)

> . . . shall employ a mechanism for monitoring and communicating throughout the organization (1) descriptions of test and field failures, (2) analyses of failure mode and root-cause failure mechanism, and (3) the status of design and/or process corrective actions and risk-mitigation decisions. This mechanism shall be accessible by the customer. . . . [The developer] shall assess the reliability of the system/product periodically throughout the life cycle. Reliability estimates from analysis, modeling and simulation, and test shall be tracked as a function of time and compared against customer reliability requirements. The implementation of corrective actions shall be verified and their effectiveness tracked. Formal reliability growth methodology shall be used where applicable . . . in order to plan, track, and project reliability improvement. . . . [The developer] shall plan and conduct activities to ensure that the design reliability requirements are met. . . . For complex systems/products, this strategy shall include reliability values to be achieved at various points during development. The verification shall be based on analysis, modeling and simulation, testing, or a mixture. . . . Testing shall be operationally realistic.

For Objective 4, to monitor and assess user reliability, the ANSI/GEIA-STD-0009 directs that RFPs mandate that proposals includes methods for which field performance can be used as feedback loops for system reliability improvement.

DTM 11-003: Improving Reliability Analysis, Planning, Tracking, and Reporting

As mentioned above, the deficiency in the reliability of fielded systems may at least be partially due to proposals that gave insufficient attention to plans for achieving reliability requirements, both initially and through testing. This issue is also addressed in DTM 11-003 (U.S. Department of Defense, 2011b, pp. 1-2), which "amplifies procedures in Reference (b) [DoD Instruction 5000.02] and is designed to improve reliability analysis, planning, tracking, and reporting." It "institutionalizes reliability planning methods and reporting requirements timed to key acquisition activities to monitor reliability growth."

DTM 11-003 stipulates that six procedures take place (pp. 3-4):

1. [Program managers] (PMs) must] formulate a comprehensive reliability and maintainability (R&M) program using an appropriate reliability growth strategy to improve R&M performance until R&M requirements are satisfied. The program will consist of engineering activities including: R&M allocations, block diagrams and predictions; failure definitions and scoring criteria; failure mode, effects and criticality analysis; maintainability and built-in test demonstrations; reliability growth testing at the system and subsystem level; and a failure reporting and corrective action system maintained through design, development, production, and sustainment. The R&M program is an integral part of the systems engineering process.
2. The lead DoD Component and the PM, or equivalent, shall prepare a preliminary Reliability, Availability, Maintainability, and Cost Rationale Report in accordance with Reference (c) [DOD Reliability, Availability, Maintainability, and Cost Rationale Report Manual, 2009] in support of the Milestone (MS) A decision. This report provides a quantitative basis for reliability requirements and improves cost estimates and program planning.
3. The Technology Development Strategy preceding MS A and the Acquisition Strategy preceding MS B and C shall specify how the sustainment characteristics of the materiel solution resulting from the analysis of alternatives and the Capability Development Document sustainment key performance parameter thresholds have been translated into R&D design requirements and contract specifications. The strategies shall also include the tasks and processes to be stated in the request for proposal that the contractor will be required to employ to demonstrate the achievement of reliability design requirements. The Test and Evaluation Strategy and the Test and Evaluation Master Plan (TEMP) shall specify how reliability will be tested and evaluated during the associated acquisition phase.
4. Reliability Growth Curves (RGC) shall reflect the reliability growth strategy and be employed to plan, illustrate and report reliability growth. A RGC shall be included in the SEP [systems engineering plan] at MS A [Milestone A], and updated in the TEMP [test and engineering master plan] beginning at MS B. RGC will be stated in a series of intermediate goals and tracked through fully integrated,

system-level test and evaluation events until the reliability threshold is achieved. If a single curve is not adequate to describe overall system reliability, curves will be provided for critical subsystems with rationale for their selection.

5. PMs and operational test agencies shall assess the reliability growth required for the system to achieve its reliability threshold during initial operational test and evaluation and report the results of that assessment to the Milestone Decision Authority at MS C.
6. Reliability growth shall be monitored and reported throughout the acquisition process. PMs shall report the status of reliability objectives and/or thresholds as part of the formal design review process, during Program Support Reviews, and during systems engineering technical reviews. RGC shall be employed to report reliability growth status at Defense Acquisition Executive System reviews.

CRITIQUE

The 2011 DOT&E Annual Report (U.S. Department of Defense, 2011a, p. iv) points out that some changes in system reliability are becoming evident:

> Sixty-five percent of FY10 TEMPs documented a reliability strategy (35 percent of those included a [reliability] growth curve), while only 20 percent of FY09 TEMPs had a documented reliability strategy. Further, three TEMPS were disapproved, citing the need for additional reliability documentation, and four other TEMPS were approved with a caveat that the next revision must include more information on the program's reliability growth strategy.

Both ANSI/GEIA-STD-0009 and DTM 11-003 serve an important purpose to help produce defense systems that (1) have more reasonable reliability requirements and (2) are more likely to meet these requirements in design and development. However, given their intended purpose, these are relatively general documents that do not provide specifics as to how some of the demands are to be met. For instance, ANSI/GEIA-STD-0009 (Information Technology Association of America, 2008, p. 2) "does not specify the details concerning how to engineer a system / product for high reliability. Nor does it mandate the methods or tools a developer would use to implement the process requirements."

The tailoring to be done will be dependent upon a "customer's funding profile, developer's internal policies and procedures and negotiations between the customer and developer" (p. 2). Proposals are to include a reliability program plan, a conceptual reliability model, an initial reliability flow-down of requirements, an initial system reliability assessment, candidate reliability trade studies, and a reliability requirements verification strategy. But there is no indication of how these activities should be carried out. How should one produce the initial reliability assessment for a system that only exists in diagrams? What does an effective design for reliability plan include? How should someone track reliability over time in development when few developmental and operationally relevant test events have taken place? How can one determine whether a test plan is adequate to take a system with a given initial reliability and improve that system's reliability through test-analyze-and-fix to the required level? How does one know when a prototype for a system is ready for operational testing?

Although the TechAmerica Engineering Bulletin Reliability Program Handbook, TA-HB-0009,[8] has been produced with the goal at least in part to answer these questions, a primary goal of this report is to assist in the provision of additional specificity as to how some of these steps should be carried out.

REFERENCES

Adolph, P., DiPetto, C.S., and Seglie, E.T. (2008). Defense Science Board task force developmental test and evaluation study results. *ITEA Journal, 29*, 215-221.

Forbes, J.A., Long, A., Lee, D.A., Essmann, W.J., and Cross, L.C. (2008). *Developing a Reliability Investment Model: Phase II—Basic, Intermediate, and Production and Support Cost Models.* LMI Government Consulting. LMI Report # HPT80T1. Available: http://www.dote.osd.mil/pub/reports/HPT80T1_Dev_a_Reliability_Investment_Model.pdf [August 2014].

Information Technology Association of America. (2008). *ANSI/GEIA-STD-0009*. Available: http://www.techstreet.com/products/1574525 [October 2014].

U.S. Department of the Army. (1996). *Pamphlet 73-2, Test and Evaluation Master Plan Procedures and Guidelines.* Available: http://acqnotes.com/Attachments/Army%20TEMP%20Procedures%20and%20Guidlines.pdf [October 2014].

U.S. Department of Defense. (2007a). *FY 2007 Annual Report.* Office of the Director of Operational Training and Development. Available: http://www.dote.osd.mil/pub/reports/FY2007/pdf/other/2007DOTEAnnualReport.pdf [January 2014].

U.S. Department of Defense. (2007b). Memorandum, *Reliability of U.S. Army Materiel Systems.* Acquisition Logistics and Technology, Assistant Secretary of the Army, Department of the Army. Available: https://dap.dau.mil/policy/Documents/Policy/Signed%20Reliability%20Memo.pdf [January 2014].

U.S. Department of Defense. (2008a). *Report of the Defense Science Board Task Force on Developmental Test and Evaluation.* Office of the Under Secretary of Defense for Acquisitions, Technology, and Logistics. Available: https://acc.dau.mil/CommunityBrowser.aspx?id=217840 [January 2014].

U.S. Department of Defense. (2008b). *Report of the Reliability Improvement Working Group.* Office of the Under Secretary of Defense for Acquisition, Technology, and Logistics. Available: http://www.acq.osd.mil/se/docs/RIWG-Report-VOL-I.pdf [January 2014].

[8]The handbook is available at http://www.techstreet.com/products/1855520 [August 2014].

U.S. Department of Defense. (2008c). Memorandum, *Implementing a Life Cycle Management Framework*. Office of the Undersecretary for Acquisition, Technology, and Logistics. Available: http://www.acq.osd.mil/log/mr/library/USD-ATL_LCM_framework_memo_31Jul08.pdf [January 2014].

U.S. Department of Defense. (2008d). *FY 2008 Annual Report*. Office of the Director of Operational Training and Development. Available: http://www.dote.osd.mil/pub/reports/FY2008/pdf/other/2008DOTEAnnualReport.pdf [January 2014].

U.S. Department of Defense. (2008e). Instruction, *Operation of the Defense Acquisition System*. Office of the Undersecretary for Acquisition, Technology, and Logistics. Available: http://www.acq.osd.mil/dpap/pdi/uid/attachments/DoDI5000-02-20081202.pdf [January 2014].

U.S. Department of Defense. (2009a). Implementation of Weapon Systems Acquisition Reform Act (WSARA) of 2009 (Public Law 111-23, May 22, 2009) October 22, 2009; Mona Lush, Special Assistant, Acquisition Initiatives, Acquisition Resources & Analysis Office of the Under Secretary of Defense for Acquisition, Technology, and Logistics.

U.S. Department of Defense. (2009b). *DoD Reliability, Availability, Maintainability-Cost (RAM-C) Report Manual*. Available: http://www.acq.osd.mil/se/docs/DoD-RAM-C-Manual.pdf [August 2014].

U.S. Department of Defense. (2010). Memorandum, *State of Reliability*. Office of the Secretary of Defense. Available: http://web.amsaa.army.mil/Documents/OSD%20Memo%20-%20State%20of%20Reliability%20-%206-30-10.pdf [January 2014].

U.S. Department of Defense. (2011a). *DOT&E FY 2011 Annual Report*. Office of the Director of Operational Test and Evaluation. Available: http://www.dote.osd.mil/pub/reports/FY2011/ [January 2014].

U.S. Department of Defense. (2011b). Memorandum, *Directive-Type Memorandum (DTM) 11-003—Reliability Analysis, Planning, Tracking, and Reporting*. The Under Secretary of Defense, Acquisition, Technology , and Logistics. Available: http://bbp.dau.mil/doc/USD-ATL%20Memo%2021Mar11%20DTM%2011-003%20-%20Reliability.pdf [January 2014].

U.S. Department of Defense. (2013). *DOT&E FY 2013 Annual Report*. Office of the Director of Operational Test and Evaluation. Available: http://www.dote.osd.mil/pub/reports/FY2013/ [January 2014].

Appendix D

Critique of MIL-HDBK-217

Anto Peter, Diganta Das, and *Michael Pecht*[1]

This paper begins with a brief history of reliability prediction of electronics and MIL-HDBK-217. It then reviews some of the specific details of MIL-HDBK-217 and its progeny and summarizes the major pitfalls of MIL-HDBK-217 and similar approaches. The effect of these shortcomings on the predictions obtained from MIL-HDBK-217 and similar methodologies are then demonstrated through a review of case studies. Lastly, this paper briefly reviews RIAC 217 Plus and identifies the shortcomings of this methodology.

HISTORY

Attempts to test and quantify the reliability of electronic components began in the 1940s during World War II. During this period, electronic tubes were the most failure-prone components used in electronic systems (McLinn, 1990; Denson, 1998). These failures led to various studies and the creation of ad hoc groups to identify ways in which the reliability of electronic systems could be improved. One of these groups concluded that in order to improve performance, the reliability of the components needed to be verified by testing before full-scale production. The specification of reliability requirements, in turn, led to a need for a method to estimate reliability before the equipment was built and tested. This step was the inception of reliability prediction for electronics. By the 1960s, spurred on

[1]The authors are at the Center for Advanced Life Cycle Engineering at the University of Maryland.

by Cold War events and the space race, reliability prediction and environmental testing became a full-blown professional discipline (Caruso, 1996).

The first dossier on reliability prediction was released by Radio Corporation of America (RCA); it was called TR-1100, Reliability Stress Analysis for Electronic Equipment. RCA was one of the major manufacturers of electronic tubes (Saleh and Marais, 2006). The report presented mathematical models for estimating component failure rates and served as the predecessor of what would become the standard and a mandatory requirement for reliability prediction in the decades to come, MIL-HDBK-217.

The methodology first used in MIL-HDBK-217 was a point estimate of the failure rate, which was estimated by fitting a line through field failure data. Soon after its introduction, all reliability predictions were based on this handbook, and all other sources of failure rates, such as those from independent experiments, gradually disappeared (Denson, 1998). The failure to use these other sources was partly due to the fact that MIL-HDBK-217 was often a contractually cited document, leaving contractors with little flexibility to use other sources.

Around the same time, a different approach to reliability estimation that focused on the physical processes by which components were failing was initiated. This approach would later be termed "physics of failure." The first symposium on this topic was sponsored by the Rome Air Development Center (RADC) and the IIT Research Institute (IITRI) in 1962.

These two—regression analysis of failure data and reliability prediction through physics of failure—seemed to be diverging, with "the system engineers devoted to the tasks of specifying, allocating, predicting, and demonstrating reliability, while the physics-of-failure engineers and scientists were devoting their efforts to identifying and modeling the physical causes of failure" (Denson, 1998, p. 3213). The result of the push toward the physics-of-failure approach was an effort to develop new models for reliability prediction for MIL-HDBK-217. These new methods were dismissed as being too complex and unrealistic. So, even though the RADC took over responsibility for preparing MIL-HDBK-217B, the physics-of-failure models were not incorporated into the revision.

As shown in Table D-1, over the course of the 1980s, MIL-HDBK-217 was updated several times, often to include newer components and more complicated, denser microcircuits. The failure rates, which were originally estimated for electronic tubes, now had to be updated to account for the complexity of devices. As a result, the MIL-HDBK-217 prediction methodology evolved to assume that times to failures were exponentially distributed and used mathematical curve fitting to arrive at a generic constant failure rate for each component type.

Other reliability models using the gate and transistor counts of microcircuits as a measure of their complexity were developed in the late 1980s

TABLE D-1 MIL-HDBK-217 Revisions and Highlights

MIL-HDBK Revision	Year and Organization in Charge	Highlights
217A	Dec 1965, Navy	Single point constant failure rate of 0.4 failures/ million hours for all monolithic ICs
217B	July 1973, Air Force Rome Labs	RCA/Boeing models simplified by Air Force to follow exponential distribution
217C	April 1979, Air Force Rome Labs	Inadequate fix for memory due to instances such as when the 4K RAM model was extrapolated to 64K, predicted MTBF = 13 sec
217D	Jan 1982, Air Force Rome Labs	No technical change in format
217E	Oct 1987, Air Force Rome Labs	No technical change in format
217F	Dec 1995, Air Force Rome Labs	CALCE, University of Maryland—Change in direction of MIL-HDBK-217 and reliability prediction recommended

(Denson and Brusius, 1989). These models were developed in support of a new MIL-HDBK-217 update. However, the gate and transistor counts eventually attained such large values that they could no longer be effectively used as a measure of complexity. This led to the development of updated reliability models that needed input parameters, such as defect density and yield of the die. But these process-related parameters were company specific and business sensitive, and hence harder to obtain. As a result, these models could not be incorporated into MIL-HDBK-217. For similar reasons, physics-of-failure-based models were also never incorporated into MIL-HDBK-217.

It was around the same time, in the 1980s, that other industries, notably, the automotive and telecommunication industries, began adapting MIL-HDBK-217 to form their own prediction methodologies and standards. The only major differences in these adaptations were that the methodologies were customized for specialized equipment under specific conditions. However, they were still based on the assumptions of the exponential distribution of failures and curve fitting to obtain a generic relationship. This approach was not surprising, because Bell Labs and Bell Communications Research (Bellcore) were the lead developers for the telecommunication reliability prediction method. Bell Labs was also one of the labs that the Navy had originally funded to investigate the reliability of electronic tubes in the 1950s; that investigation culminated in the drafting of MIL-HDBK-217. Hence, with the automotive and telecommunication industries adopting methodologies similar to MIL-HDBK-217, the handbook's practices proliferated in the commercial sector. While this was happening, there were also several researchers who had experimental data to show that the handbook-based methodologies were fundamentally flawed in their assumptions. However, these were often either explained away as being anomalies or labeled as invalid.

By the 1990s, the handbooks were struggling to keep up with new components and technological advancements. In 1994, there was another major development in the initiation of the U.S. Military Specification and Standards Reform (MSSR). The MSSR identified MIL-HDBK-217 as the only standard that "required priority action as it was identified as a barrier to commercial processes as well as major cost drivers in defense acquisitions" (Denson, 1998, p. 3214). Despite this, no final model was developed to supplement or replace MIL-HDBK-217 in the 1990s. Instead, a final revision was made to the handbook in the form of MIL-HDBK-217F in 1995.

At this stage, the handbook-based methodologies were already outdated in their selection of components and the nature of failures considered. In the 1990s, the electronic systems were vastly more complicated and sophisticated than they had been in the 1960s, when the handbook was developed. Failure rates were no longer determined by components, but rather by system-level factors such as manufacturing, design, and software interfaces. Based on the new understanding of the critical failure mechanisms in systems and the physics underlying failures, MIL-HDBK-217 was found to be completely incapable of being applied to systems to predict system reliability.

Moving forward into the 2000s and up through 2013, methodologies based on MIL-HDBK-217 were still being used in the industry to predict reliability and provide such metrics as mean time to failure and mean time between failures (MTTF and MTBF). These metrics are still used as estimates of reliability, even though both the methodologies and the database of failure rates used to evaluate the metrics are outdated.

The typical feature size when MIL-HDBK-217 was last updated was of the order of 500 nm, while commercially available electronic packages today have feature sizes of 22 nm (e.g., Intel Core i7 processor). Furthermore, many components, both active and passive, such as niobium capacitors and insulated gate bipolar transistors (IGBTs), which are now common, had not been invented at the time of the last MIL-HDBK-217 revision. Needless to say, the components and their use conditions, the failure modes and mechanisms, and the failure rates for today's systems are vastly different from the components for which MIL-HDBK-217 was developed. Hence, the continued application of these handbook methodologies by the industry is misguided and misleading to customers and designers alike. The solution is not to develop an updated handbook like the RIAC 217 Plus, but rather to concede that reliability cannot be predicted by deterministic models. The use of methodologies based on MIL-HDBK-217 has proven to be detrimental to the reliability engineering community as a whole.

MIL-HDBK-217 AND ITS PROGENY

In this section, we review the different standards and reliability prediction methodologies, comparing them to the latest draft of IEEE 1413.1, Guide for Developing and Assessing Reliability Predictions Based on IEEE Standard 1413 (Standards Committee of the IEEE Reliability, 2010). Most handbook-based prediction methodologies can be traced back to MIL-HDBK-217 and are treated as its progeny. As mentioned above, MIL-HDBK-217 was based on curve fitting a mathematical model to historical field failure data to determine the constant failure rate of parts. Its progeny also use similar prediction methods, which are based purely on fitting a curve through field or test failure data. These methodologies, much like MIL-HDBK-217, use some form of a constant failure rate model: they do not consider actual failure modes or mechanisms. Hence, these methodologies are only applicable in cases where systems or components exhibit relatively constant failure rates. Table D-2 lists some of the standards and prediction methodologies that are considered to be the progeny of MIL-HDBK-217.

TABLE D-2 MIL-HDBK-217-Related Reliability Prediction Methodologies and Applications

Procedural Method	Applications	Status
MIL-HDBK-217	Military	Active
Telcordia SR-332	Telecom	Active
CNET	Ground Military	Canceled
RDF-93 and 2000	Civil Equipment	Active
SAE Reliability Prediction	Automotive	Canceled
British Telecom HRD-5	Telecom	Canceled
Siemens SN29500	Siemens Products	Canceled
NTT Procedure	Commercial and Military	Canceled
PRISM	Aeronautic and Military	Active
RIAC 217Plus	Aeronautic and Military	Active
FIDES	Aeronautic and Military	Active

In most cases, the failure rate relationship that is used by these handbook techniques (adapted from MIL-HDBK-217) takes the form of $\lambda_p = f(\lambda_G, \pi_i)$ where λ_p is the calculated constant part failure rate; λ_G is a constant part failure rate (also known as base failure rate), which is provided by the handbook; and π_i is a set of adjustment factors for the assumed constant failure rates. All of these handbook methods either provide a constant failure rate or a method to calculate it. The handbook methods that calculate constant failure rates use one or more multiplicative adjustment factors (which may include factors for part quality, temperature, design, or environment) to modify a given constant base failure rate.

The constant failure rates in the handbooks are obtained by performing a linear regression analysis on the field data. The aim of the regression analysis is to quantify the expected theoretical relationship between the constant part failure rate and the independent variables. The first step in the analysis is to examine the correlation matrix for all variables, showing the correlation between the dependent variable (the constant failure rate) and each independent variable. The independent variables used in the regression analysis typically include such factors as the device type, package type, screening level, ambient temperature, and application stresses. The second step is to apply stepwise multiple linear regressions to the data, which express the constant failure rate as a function of the relevant independent variables and their respective coefficients. This is the step that involves the evaluation of the above π factors. The constant failure rate is then calculated using the regression formula and the input parameters.

The regression analysis does not ignore data entries that lack essential information, because the scarcity of data necessitates that all available data be used. To accommodate such data entries in the regression analysis, a separate "missing" category may be constructed for each potential factor when the required information is not available. A regression factor can be calculated for each "missing" category, considering it a unique operational condition. If the coefficient for the unknown category is significantly smaller than the next lower category or larger than the next higher category, then that factor in question cannot be quantified by the available data, and additional data are required before the factor can be fully evaluated (Standards Committee of the IEEE Reliability, 2010).

A constant failure rate model for non-operating conditions can be extrapolated by eliminating all operation-related stresses from the handbook prediction models, such as temperature rise or electrical stress ratio. Following the problems related to the use of missing data, using handbooks such as MIL-HDBK-217 to calculate constant non-operating failure rates is an extrapolation of the empirical relationship of the source field data beyond the range in which it was gathered. In other words, the

TABLE D-3 Constant Failure Rate Calculations of Handbook Methodologies

Method	Failure Rate for Microelectronic Devices
MIL-HDBK-217 (parts count)	$\lambda = \lambda_G \Pi_Q \Pi_L$
MIL-HDBK-217 (parts stress)	$\lambda = \Pi_Q (C_1 \Pi_T \Pi_V + C_2 \Pi_E) \Pi_L$
SAE PREL	$\lambda_p = \lambda_b \Pi_Q \Pi_S \Pi_T \Pi_E$
Telcordia SR-332	$\lambda = \lambda_G \Pi_Q \Pi_S \Pi_T$
British Telecom HRD-5	$\lambda = \lambda_b \Pi_T \Pi_Q \Pi_E$
PRISM	$\lambda_p = \lambda_{IA} (\Pi_P + \Pi_D + \Pi_M + \Pi_S) + \lambda_{SW} + \lambda_W$
CNET (simplified)	$\lambda = \Pi_Q \lambda_A$
CNET (stress model)	$\lambda_p = (C_1 \Pi_t \Pi_T \Pi_V + C_2 \Pi_B \Pi_E \Pi_\sigma) \Pi_L \Pi_Q$
Siemens SN29500	$\lambda = \lambda_b \Pi_U \Pi_T$

NOTES: Π_L is a learning factor, Π_T is the temperature factor, Π_E is the environment factor, Π_Q is the quality factor, C_1 is the die complexity, and C_2 is the package complexity. For additional details, see U.S. Department of Defense (1991).

MIL-HDBK-217-based constant failure rates are not applicable to failures related to storage and handling.

Table D-3 lists the typical constant failure rate calculations that are used by various handbook methodologies. Most of these methods have a form that is very similar to MIL-HDBK-217, despite the modifications that have been made for environmental and application-specific loading conditions. Some of these methodologies are described briefly in the following sections.

MIL-HDBK-217F

MIL-HDBK-217 provides two constant failure rate prediction methods: parts count and parts stress. The MIL-HDBK-217F parts stress method provides constant failure rate models based on curve-fitting the empirical data obtained from field operation and testing. The models have a constant base failure rate modified by environmental, temperature, electrical stress, quality, and other factors. Both methods are based on $\lambda_p = f(\lambda_G, \pi_i)$, but the parts stress method assumes there are no modifiers to the general constant failure rate. The MIL-HDBK-217 methodology only provides results for parts, not for equipment or systems.

TELCORDIA SR-332

Telcordia SR-332 is a reliability prediction methodology developed by Bell Communications Research (or Bellcore) primarily for telecommunications companies (Telcordia Technologies, 2001). The most recent revision of the methodology is Issue 3, dated January 2011. The stated purpose of Telcordia SR-332 is "to document the recommended methods for predicting device and unit hardware reliability (and also) for predicting serial system hardware reliability" (Telcordia Technologies, 2001, p. 1-1). The methodology is based on empirical statistical modeling of commercial telecommunication systems whose physical design, manufacture, installation, and reliability assurance practices meet the appropriate Telcordia (or equivalent) generic and system-specific requirements. In general, Telcordia SR-332 adapts the equations in MIL-HDBK-217 to represent the conditions that telecommunication equipment experience in the field. Results are provided as a constant failure rate, and the handbook provides the upper 90 percent confidence-level point estimate for the constant failure rate.

The main concepts in MIL-HDBK-217 and Telcordia SR-332 are similar, but Telcordia SR-332 also has the ability to incorporate burn-in, field, and laboratory test data for a Bayesian analytical approach that incorporates both prior information and observed data to generate an updated posterior distribution. For example, Telcordia SR-332 contains a table of the "first-year multiplier" (Telcordia Technologies, 2001, p. 2-2), which is the predicted ratio of the number of failures of a part in its first year of operation in the field to the number of failures of the part in another year of (steady state) operation. This table in the SR-332 contains the first-year multiplier for each value of the part device burn-in time in the factory. The part's total burn-in time is the sum of the burn-in time at the part, unit, and system levels.

PRISM

PRISM is a reliability assessment method developed by the Reliability Analysis Center (RAC) (Reliability Assessment Center, 2001). The method is available only as software, and the most recent version of the software is Version 1.5, released in May 2003. PRISM combines the empirical data of users with a built-in database using Bayesian techniques. In this technique, new data are combined using a weighted average method, but there is no new regression analysis. PRISM includes some nonpart factors such as interface, software, and mechanical problems.

PRISM calculates assembly- and system-level constant failure rates in accordance with similarity analysis, which is an assessment method that compares the actual life-cycle characteristics of a system with predefined process grading criteria, from which an estimated constant failure rate is obtained. The component models used in PRISM are called RACRates™ models and are based on historical field data acquired from a variety of sources over time and under various undefined levels of statistical control and verification.

Unlike the other handbook constant failure rate models, the RACRates™ models do not have a separate factor for part quality level. Quality level is implicitly accounted for by a method known as process grading. Process grades address factors such as design, manufacturing, part procurement, and system management, which are intended to capture the extent to which measures have been taken to minimize the occurrence of system failures.

The RACRates™ models consider separately the following five contributions to the total component constant failure rate: (1) operating conditions, (2) non-operating conditions, (3) temperature cycling, (4) solder joint reliability, and (5) electrical overstress (EOS). Solder joint failures are also combined with other failures in the model, without consideration of the board material or solder material. These five factors are not independent: for example, solder joint failures depend on the temperature cycling parameters. A constant failure rate is calculated for solder joint reliability, although solder joint failures are primarily wear-out failure mechanisms due to cyclic fatigue.

PRISM calculates non-operating constant failure rates with several assumptions. The daily or seasonal temperature cycling high and low values that are assumed to occur during storage or dormancy represent the largest contribution to the non-operating constant failure rate value. The contribution of solder joints to the non-operating constant failure rate value is represented by reducing the internal part temperature rise to zero for each part in the system. Lastly, the contribution of the probability of electrical overstress (EOS) or electrostatic discharge (ESD) is represented by the assumption that the EOS constant failure rate is independent of the duty cycle. This assumption accounts for parts in storage affected by this EOS or ESD due to handling and transportation.

FIDES GUIDE

The FIDES methodology was developed under the supervision of Délégation Générale pour l'Armement, specifically for the French Ministry of Defense. The methodology was formed by French industrialists from the fields of aeronautics and defense. It was compiled by the following organizations: AIRBUS France, Eurocopter, GIAT Industries, MBDA France, Thales Airborne Systems, Thales Avionics, Thales Research & Technology, and Thales Underwater Systems. The FIDES Guide aims "to enable a realistic assessment of the reliability of electronic equipment, including systems operating in severe environments (defense systems, aeronautics, industrial electronics, and transport). The FIDES Guide also aims to provide a concrete tool to develop and control reliability" (FIDES Group, 2009).

The FIDES Guide contains two parts: a reliability prediction model and a reliability process control and audit guide. The FIDES Guide provides models for electrical, electronic, and electromechanical components. These prediction models take into account the electrical, mechanical, and thermal overstresses. These models also account for "failures linked to . . . development, production, field operation and maintenance" (FIDES Group, 2009). The reliability process control guide addresses the procedures and organizations throughout the life cycle, but does not go into the use of the components themselves. The audit guide is a generic procedure that audits a company using three questions as a basis to "measure its capability to build reliable systems, quantify the process factors used in the calculation models, and identify actions for improvement" (FIDES Group, 2009).

NON-OPERATING CONSTANT FAILURE RATE PREDICTIONS

MIL-HDBK-217 does not have specific methods or data related to the non-operational failure of electronic parts and systems, although several different methods to estimate them were proposed in the 1970s and 1980s. The first methods used multiplicative factors based on the operating constant failure rates obtained using other handbook methods. The reported values of such multiplicative factors are 0.03 or 0.1. The first value of 0.03 was obtained from an unpublished study of satellite clock failure data from 23 failures. The value of 0.1 is based on a RADC study from 1980. RAC followed up the efforts with the RADC-TR-85-91 method. This method was described as being equivalent to MIL-HDBK-217 for non-operating conditions, and it contained the same number of environmental factors and the same type of quality factors as the then-current MIL-HDBK-217. Some other non-operating constant failure rate tables from the 1970s and 1980s include the MIRADCOM Report LC-78-1, RADC-TR-73-248, and NONOP-1.

IEEE 1413 AND COMPARISON OF RELIABILITY PREDICTION METHODOLOGIES

The IEEE Standard 1413, IEEE Standard Methodology for Reliability Prediction and Assessment for Electronic Systems and Equipment (IEEE Standards Association, 2010), provides a framework for reliability prediction procedures for electronic equipment at all levels. It focuses on hardware reliability prediction methodologies, and specifically excludes software reliability, availability and maintainability, human reliability, and proprietary reliability prediction data and methodologies. IEEE 1413.1, Guide for Selecting and Using Reliability Predictions Based on IEEE1413 (IEEE Standards Association, 2010) aids in the selection and use of reliability prediction methodologies that satisfy IEEE 1413. Table D-4 shows a comparison of some of the handbook-based reliability prediction methodologies based on the criteria in IEEE 1413 and 1413.1.

Though only five of the many failure prediction methodologies have been analyzed, they are representative of the other constant failure-rate-based techniques. There have been several publications that assess other similar aspects of prediction methodologies. Examples include O'Connor (1985a, 1985b, 1988, 1990), O'Connor and Harris (1986), Bhagat (1989), Leonard (1987, 1988), Wong (1990, 1993, 1989), Bowles (1992), Leonard and Pecht (1993), Nash (1993), Hallberg (1994), and Lall et al. (1997).

These methodologies do not identify the root causes, failure modes, and failure mechanisms. Therefore, these techniques offer limited insight into the real reliability issues and could potentially misguide efforts to design for reliability, as is demonstrated in Cushing et al. (1996), Hallberg (1987, 1991), Pease (1991), Watson (1992), Pecht and Nash (1994), and Knowles (1993). The following sections will review some of the major shortcomings of handbook-based methodologies, and also present case studies highlighting the inconsistencies and inaccuracies of these approaches.

SHORTCOMINGS OF MIL-HDBK-217 AND ITS PROGENY

MIL-HDBK-217 has several shortcomings, and it has been critiqued extensively since the early 1960s. Some of the initial arguments and results contradicting the handbook methodologies were refuted as being fraudulent or sourced from manipulated data (see McLinn, 1990). However, by the early 1990s, it was agreed that MIL-HDBK-217 was severely limited in its capabilities, as far as reliability prediction was concerned (Pecht and Nash, 1994). One of the main drawbacks of MIL-HDBK-217 was that the predictions were based purely on "simple heuristics," as opposed to engineering design principles and physics of failure. The handbook could not even account for different loading conditions (Jais et al., 2013). Furthermore, because the handbook was focused mainly on component-level analyses, it could only address a fraction of overall system failure rates. In addition to these issues, if MIL-HDBK-217 was only used for arriving at a rough estimate of reliability of a component, then it would need to be constantly updated with the newer technologies, but this was never the case.

TABLE D-4 Comparison of Reliability Prediction Methodologies

Questions for Comparison	MIL-HDBK-217F	Telcordia SR-332	217Plus	PRISM	FIDES
Does the methodology identify the sources used to develop the prediction methodology and describe the extent to which the source is known?	Yes	Yes	Yes	Yes	Yes
Are assumptions used to conduct the prediction according to the methodology identified, including those used for the unknown data?	Yes	Yes	Yes	Yes	Yes (must pay for modeling software).
Are sources of uncertainty in the prediction results identified?	No	Yes	Yes	No	Yes
Are limitations of the prediction results identified?	Yes	Yes	Yes	Yes	Yes
Are failure modes identified?	No	No	No	No	Yes, the failure mode profile varies with the life profile.
Are failure mechanisms identified?	No	No	No	No	Yes
Are confidence levels for the prediction results identified?	No	Yes	Yes	No	No
Does the methodology account for life-cycle environmental conditions, including those encountered during (a) product usage (including power and voltage conditions), (b) packaging, (c) handling, (d) storage, (e) transportation, and (f) maintenance conditions?	No. It does not consider the different aspects of environment. There is a temperature factor π_T and an environment factor π_E in the prediction equation.	Yes, for normal use life of the product from early life to steady-state operation over the normal product life.	No	No	Yes. It considers all of the life-cycle environmental conditions.

Does the methodology account for materials, geometry, and architectures that comprise the parts?	No	No	No	No	Yes, when relevant materials, geometry and such are considered in each part model.
Does the methodology account for part quality?	Quality levels are derived from specific part-dependent data and the number of the manufacturer screens the part goes through.	Four quality levels that are based on generalities regarding the origin and screening of parts.	Quality is accounted for in the part quality process grading factor.	Part quality level is implicitly addressed by process grading factors and the growth factor, p_G.	Yes
Does the methodology allow incorporation of reliability data and experience?	No	Yes, through Bayesian method of weighted averaging.	Yes, through Bayesian method of weighed averaging	Yes, through Bayesian method of weighted averaging.	Yes. This can be done independently of the prediction methodology used.
Input data required for the analysis	Information on part count and operational conditions (e.g., temperature, voltage; specifics depend on the handbook used).				
Other requirements for performing the analysis	Effort required is relatively small for using the handbook method and is limited to obtaining the handbook.				
What is the coverage of electronic parts?	Extensive	Extensive	Extensive	Extensive	Extensive
What failure probability distributions are supported?	Exponential	Exponential	Exponential	Exponential	Phase Contributions

SOURCE: Adapted from IEEE 1413.1.

Incorrect Assumption of Constant Failure Rates

MIL-HDBK-217 assumes that the failure rates of all electronic components are constant, regardless of the nature of actual stresses that the component or system experiences. This assumption was first made based on statistical consideration of failure data independent of the cause or nature of failures. Since then, significant developments have been made in the understanding of the physics of failure, failure modes, and failure mechanisms.

It is now understood that failure rates, and, more specifically, the hazard rates (or instantaneous failure rates), vary with time. Studies have shown that the hazard rates for electronic components, such as transistors, that are subjected to high temperatures and high electric fields, under common failure mechanisms, show an increasing hazard rate with time (Li et al., 2008; Patil et al., 2009). At the same time, in components and systems with manufacturing defects, failures may manifest themselves early on, and as these parts fail, they get screened. Therefore, a decreasing failure rate may be observed initially in the life of a product. Hence, it is safe to say that in the life cycle of an electronic component or system, the failure rate is constantly varying.

McLinn (1990) and Bowles (2002) describe the history, math, and flawed reasoning behind constant failure rate assumptions. Epstein and Sobel (1954) provide a historical review of some of the first applications of the exponential distribution to model mortality in actuarial studies for the insurance industry in the early 1950s. Since exponential distributions are associated with constant failure rates, which help simplify calculations, they were adopted by the reliability engineering community. Through subsequent widespread usage, "the constant failure rate model, right or wrong, became the 'reliability paradigm'" (McLinn, 1990, p. 237). McLinn notes how once this paradigm was adopted, its practitioners, based on their common beliefs, "became committed more to the propagation of the paradigm, than the accuracy of the paradigm itself" (p. 239).

By the end of the 1950s, and in the early 1960s, more test data were obtained from experiments. The data seemed to indicate that electronic systems at that time had decreasing failure rates (Milligan, 1961; Pettinato and McLaughlin, 1961). However, the natural tendency of the proponents of the constant failure rate model was to explain away these results as anomalies as opposed to providing a "fuller explanation" (McLinn, 1990, p. 240). Concepts such as inverse burn-in or endless burn-in (Bezat and Montague, 1979; McLinn, 1989) and mysterious unexplained causes were used to dismiss anomalies.

Proponents of the constant failure rate model believed that the hazard rates or instantaneous failure rates of electronic systems would follow a bathtub curve—with an initial region (called infant mortality) of decreasing failure rate when failures due to manufacturing defects would be weeded out. This stage would then be followed by a region of a constant failure rate, and toward the end of the life cycle the failure rate would increase due to wear-out mechanisms. This theory would help reconcile both the decreasing failure rate and the constant failure rate model. When Peck (1971) published data from semiconductor testing, he observed a decreasing failure rate trend that lasted for many thousands of hours of operation. This was said to have been caused by "freaks." It was later explained as being an extended infant mortality rate.

Bellcore and SAE created two standards using a prediction methodology based on constant failure rate, but they subsequently adjusted their techniques to account for this phenomenon (of decreasing failure rates lasting several thousand hours) by increasing the infant mortality region to 10,000 hours and 100,000 hours, respectively. However, the bathtub curve theory was further challenged by Wong (1981) with his work describing the demise of the bathtub curve. Claims were made by constant failure rate proponents suggesting that data challenging the bathtub curve and constant failure rate models were fraudulently manipulated (Ryerson, 1982). These allegations were merely asserted with no supporting analysis or explanation. In order to reconcile some of the results from contemporary publications, the roller coaster curve—which was essentially a modified bathtub curve—was introduced by Wong and Lindstrom (1989). McLinn (1990, p. 239) noted that the arguments and modifications made by the constant failure rate proponents "were not always based on science or logic . . . but may be unconsciously based on a desire to adhere to the old and familiar models." Figure D-1 depicts the bathtub curve and the roller coaster curve.

There has been much debate about the suitability of the constant failure rate assumption for modeling component reliability. This methodology has been controversial in terms of assessing reliability during design: see, for

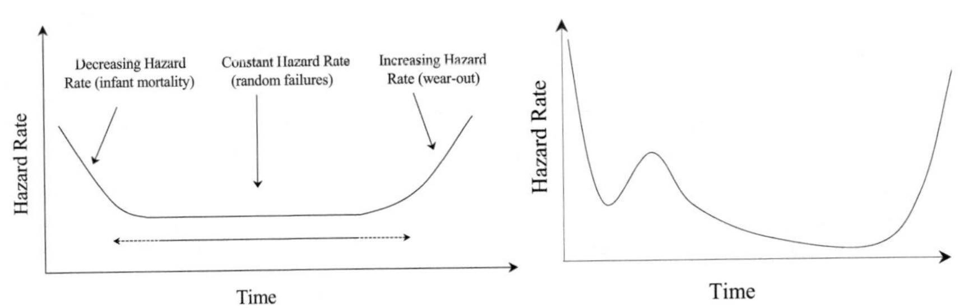

FIGURE D-1 The bathtub curve (left) and the roller coaster curve (right). See text for discussion.

example, Blanks (1980), Bar-Cohen (1988), Coleman (1992), and Cushing et al. (1993). The mathematical perspective has been discussed in detail in Bowles (2002). Thus, a complete understanding of how the constant failure rates are evaluated, along with the implicit assumptions, is vital to interpreting both reliability predictions and future design. It is important to remember that the constant failure rate models used in some of the handbooks are calculated by performing a linear regression analysis on the field failure data or generic test data. These data and the constant failure rates are not representative of the actual failure rates that a system might experience in the field (unless the environmental and loading conditions are static and the same for all devices). Because a device might see several different types of stresses and environmental conditions, it could be degrading in multiple ways. Hence, the lifetime of an electronic compound or device can be approximated to be a combination of several different failure mechanisms and modes, each having its own distribution, as shown in Figure D-2.

Furthermore, this degradation is nondeterministic, so the product will have differing failure rates throughout its life. It would be impossible to capture this behavior in a constant failure rate model. Therefore, all methodologies based on the assumption of a constant failure rate are fundamentally flawed and cannot be used to predict reliability in the field.

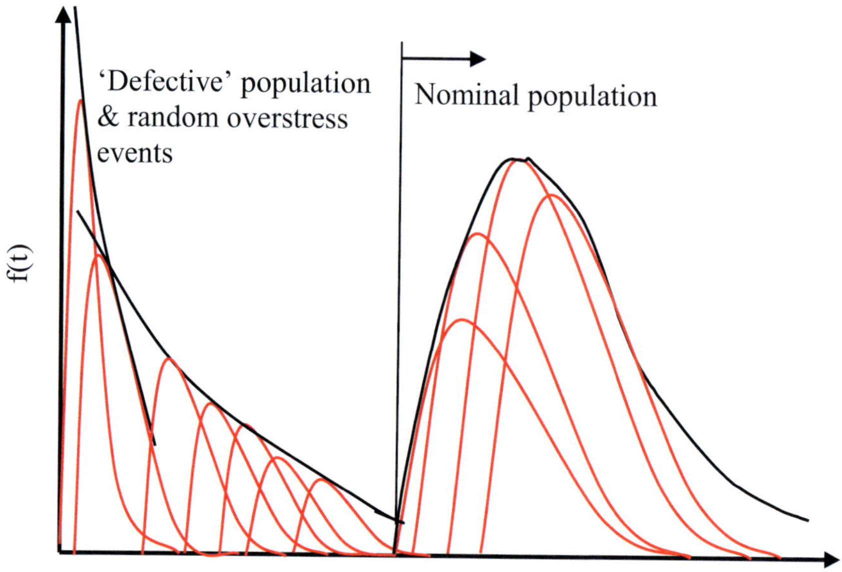

FIGURE D-2 The physics-of-failure perspective on the bathtub curve and failure rates. See text for discussion.

Lack of Consideration of Root Causes of Failures, Failure Modes, and Failure Mechanisms

After the assumption of constant failure rates, the lack of any consideration of root causes of failures and failure modes and mechanisms is another major drawback of MIL-HDBK-217 and other similar approaches. Without knowledge or understanding of the site, root cause, or mechanism of failures, the load and environment history, the materials, and the geometries, the calculated failure rate is meaningless in the context of reliability prediction. As noted above, not only does this undermine reliability assessment in general, it also obstructs product design and process improvement.

Cushing et al. (1993) note that there are two major consequences of using MIL-HDBK-217. First, this prediction methodology "does not give the designer or manufacturer any insight into, or control over, the actual causes of failure since the cause-and-effect relationships impacting reliability are not captured. Yet, the failure rate obtained is often used as a reverse engineering tool to meet reliability goals." At the same time, "MIL-HDBK-217 does not address the design & usage parameters that greatly influence reliability, which results in an inability to tailor a MIL-HDBK-217 prediction using these key parameters" (Cushing et al., 1993, p. 542).

In countries such as Japan (see Kelly et al., 1995), Singapore, and Taiwan, where the focus is on product improvement, physics of failure is the only approach used for reliability assessment. In stark contrast, in the United States and Europe, focus on "quantification of reliability and device failure rate prediction has been more common" (Cushing et al., 1993, p. 542). This approach has turned reliability assessment into a numbers game, with greater importance being given to the MTBF value and the failure rate than to the cause of failure. The reason most often cited for this rejection of physics-of-failure-based approaches in favor of simplistic mathematical regression analysis was the complicated and sophisticated nature of physics-of-failure models. In hindsight, this rejection of physics-based models, without completely evaluating the merits of the approach and without having any foresight, was poor engineering practice.

Though MIL-HDBK-217 provides inadequate design guidance, it has often been used in the design of boards, circuit cards, and other assemblies. Research sponsored by the U.S. Army (Pecht et al., 1992) and the National Institute of Standards and Technology (NIST) (Kopanski et al., 1991) explored an example of design misguidance resulting from device failure rate prediction methodologies concerning the relationship between thermal stresses and microelectronic failure mechanisms. In this case, MIL-HDBK-217 would clearly not be able to distinguish between the two separate failure mechanisms. Results from another independent study by Boeing

(Leonard, 1991) corroborated these findings. The MIL-HDBK-217-based methodologies also cannot be used for comparison and evaluation of competing designs. They cannot provide accurate comparisons or even a specification of accuracy.

Physics of failure, in contrast "is an approach to design, reliability assessment, testing, screening and evaluating stress margins by employing knowledge of root-cause failure processes to prevent product failures through robust design and manufacturing practices" (Lall and Pecht, 1993, p. 1170). The physics-of-failure approach to reliability involves many steps, including: (1) identifying potential failure mechanisms, failure modes, and failure sites; (2) identifying appropriate failure models and their input parameters; (3) determining the variability of each design parameter when possible; (4) computing the effective reliability function; and (5) accepting the design, if the estimated time-dependent reliability function meets or exceeds the required value over the required time period.

Table D-5 compares several aspects of MIL-HDBK-217 with those of the physics-of-failure approach.

Several physics-of-failure-based models have been developed for different types of materials, at different levels of electronic packaging (chip, component, board), and under different loading conditions (vibration, chemical, electrical). Though it would be impossible to list or review all of them, many models have been discussed in the physics-of-failure tutorial series in *IEEE Transactions on Reliability*. Examples include Dasgupta and Pecht (1991), Dasgupta and Hu (1992a), Dasgupta and Hu (1992b), Dasgupta (1993), Dasgupta and Haslach (1993), Engel (1993), Li and Dasgupta (1993), Al-Sheikhly and Christou (1994), Li and Dasgupta (1994), Rudra and Jennings (1994), Young and Christou (1994), and Diaz et al. (1995).

The physics-of-failure models do have some limitations as well. The results obtained from these models will have a certain degree of uncertainty and errors associated with them, which can be partly mitigated by calibrating them with accelerated testing. The physics-of-failure methods may also be limited in their ability to combine the results of the same model for multiple stress conditions or their ability to aggregate the failure prediction results from individual failure modes to a complex system with multiple competing and common cause failure modes. However, there are recognized methods to address these issues, with continuing research promising improvements; for details, see Asher and Feingold (1984), Montgomery and Runger (1994), Shetty et al. (2002), Mishra et al. (2004), and Ramakrishnan and Pecht (2003).

Despite the shortcomings of the physics-of failure-approach, it is more rigorous and complete, and, hence, it is scientifically superior to the constant failure rate models. The constant failure rate reliability predictions have little relevance to the actual reliability of an electronic system in the

TABLE D-5 A Comparison Between the MIL-HDBK-217 and Physics-of-Failure Approaches

Issue	MIL-HDBK-217	Physics-of-Failure Approach
Model Development	Models cannot provide accurate design or manufacturing guidance since they were developed from assumed constant failure-rate data, not root-cause, time-to-failure data. A proponent stated: "Therefore, because of the fragmented nature of the data and the fact that it is often necessary to interpolate or extrapolate from available data when developing new models, no statistical confidence intervals should be associated with the overall model results" (Morris, 1990).	Models based on science/engineering first principles. Models can support deterministic or probabilistic applications.
Device Design Modeling	The MIL-HDBK-217 assumption of perfect designs is not substantiated due to lack of root-cause analysis of field failures. MIL-HDBK-217 models do not identify wearout issues.	Models for root-cause failure mechanisms allow explicit consideration of the impact that design, manufacturing, and operation have on reliability.
Device Defect Modeling	Models cannot be used to (1) consider explicitly the impact of manufacturing variation on reliability, or (2) determine what constitutes a defect or how to screen/inspect defects.	Failure mechanism models can be used to (1) relate manufacturing variation to reliability, and (2) determine what constitutes a defect and how to screen/inspect.
Device Screening	MIL-HDBK-217 promotes and encourages screening without recognition of potential failure mechanisms.	Provides a scientific basis for determining the effectiveness of particular screens or inspections.
Device Coverage	Does not cover new devices for approximately the first 5–8 years. Some devices, such as connectors, were not updated for more than 20 years. Developing and maintaining current design reliability models for devices is an impossible task.	Generally applicable—applies to both existing and new devices—since failure mechanisms are modeled, not devices. Thirty years of reliability physics research has produced and continues to produce peer-reviewed models for the key failure mechanisms applicable to electronic equipment. Automated computer tools exist for printed wiring boards and microelectronic devices.

continued

TABLE D-5 Continued

Issue	MIL-HDBK-217	Physics-of-Failure Approach
Use of Arrhenius Model	Indicates to designers that steady-state temperature is the primary stress that designers can reduce to improve reliability. MIL-HDBK-217 models will not accept explicit temperature change inputs. MIL-HDBK-217 lumps different acceleration models from various failure mechanisms together, which is unsound.	The Arrhenius model is used to model the relationships between steady-state temperature and mean time-to-failure for each failure mechanism, as applicable. In addition, stresses due to temperature change, temperature rate of change, and spatial temperature gradients are considered, as applicable.
Operating Temperature	Explicitly considers only steady-state temperature. The effect of steady-state temperature is inaccurate because it is not based on root-cause, time-to-failure data.	The appropriate temperature dependence of each failure mechanism is explicitly considered. Reliability is frequently more sensitive to temperature cycling, provided that adequate margins are given against temperature extremes (see Pecht et al., 1992).
Operational Temperature Cycling	Does not support explicit consideration of the impact of temperature cycling on reliability. No way of superposing the effects of temperature cycling and vibration.	Explicitly considers all stresses, including steady-state temperature, temperature change, temperature rate of change, and spatial temperature gradients, as applicable to each root-cause failure mechanism.
Input Data Required	Does not model critical failure contributors, such as materials architectures, and realistic operating stresses. Minimal data in, minimal data out.	Information on materials, architectures, and operating stresses—the things that contribute to failures. This information is accessible from the design and manufacturing processes of leading electronics companies.
Output Data	Output is typically a (constant) failure rate 'λ'. A proponent stated: "MIL-HDBK-217 is not intended to predict field reliability and, in general, does not do a very good job in an absolute sense" (Morris, 1990).	Provides insight to designers on the impact of materials, architectures, loading, and associated variation. Predicts the time-to-failure (as a distribution) and failure sites for key failure mechanisms in a device or assembly. These failure times and sites can be ranked. This approach supports either deterministic or probabilistic treatment.

TABLE D-5 Continued

Issue	MIL-HDBK-217	Physics-of-Failure Approach
DoD/Industry Acceptance	Mandated by government; 30-year record of discontent. Not part of the U.S. Air Force Avionics Integrity Program (AVIP). No longer supported by senior U.S. Army leaders.	Represents the best practices of industry.
Coordination	Models have never been submitted to appropriate engineering societies and technical journals for formal peer review.	Models for root-cause failure mechanisms undergo continuous peer review by leading experts. New software and documentation are coordinated with the various DoD branches and other entities.
Relative Cost of Analysis	Cost is high compared with value added. Can misguide efforts to design reliable electronic equipment.	Intent is to focus on root-cause failure mechanisms and sites, which is central to good design and manufacturing. Acquisition flexible, so costs are flexible. The approach can result in reduced life-cycle costs due to higher initial and final reliabilities, reduced probability of failing tests, reductions in hidden factory, and reduced support costs.

SOURCE: Cushing et al. (1993). Reprinted with permission.

field. The weaknesses of the physics-of-failure approach can mostly be attributed to the lack of knowledge of the exact usage environment and loading conditions that a device might experience and the stochastic nature of the degradation process. However, with the availability of various sensors for data collection and data transmission, this gap in knowledge is being overcome. The weaknesses of the physics-of-failure approach can also be overcome by augmenting it with prognostic and health management approaches, such as the one shown in Figure D-3 (see Pecht and Gu, 2009).

The process based on prognostics and health management does not predict reliability, but it does provide a reliability assessment based on in-situ monitoring of certain environmental or performance parameters. This process combines the strengths of the physics-of-failure approach with live monitoring of the environment and operational loading conditions.

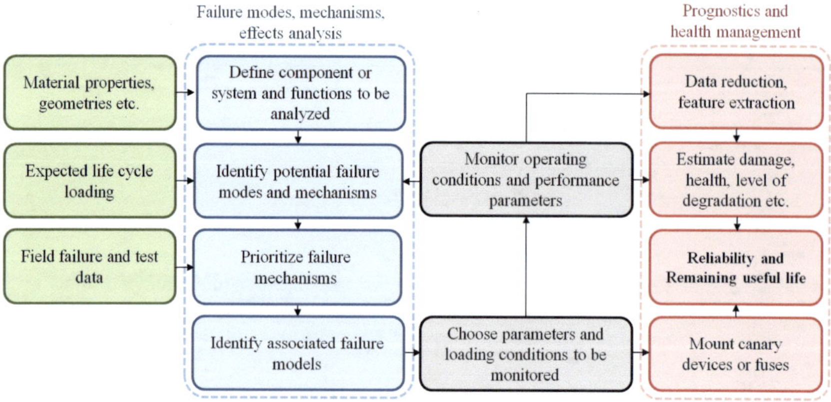

FIGURE D-3 Physics-of-failure-based prognostics and health management approach. For details, see Pecht and Gu (2009, p. 315).
SOURCE: Adapted from Gu and Pecht (2007). Reprinted with permission.

Inadequacy for System-Level Considerations

The MIL-HDBK-217 methodology is predominantly focused on component-level reliability prediction, not system-level reliability. This focus may not have been unreasonable in the 1960s and 1970s, when components had higher failure rates and electronic systems were less complex than they are today. However, as Denson (1988) notes, an increase in system complexity and component quality has resulted in a shift of system-failure causes away from components toward system-level factors, including the manufacturing, design, system requirements, and interface. Aspects such as human factors or operator errors, faulty maintenance or installation, equipment-to-equipment interaction, and software reliability also have a significant impact on the reliability of a system, and hence these factors should be accounted for in the reliability prediction (Halverson and Ozdes, 1992; Jensen, 1995).

Historically, none of these factors have been addressed by MIL-HDBK-217 (Denson, 1998). As a result, only a fraction of a total system failure rate is accounted for using handbook-based techniques: see the results from surveys conducted by Denson (1998) and Pecht and Nash (1994), shown in Table D-6. In order to estimate system-level reliability, MIL-HDBK-217 suggests that the individual reliabilities of components either be added or multiplied with other correction factors, all with the overarching assumption of constant failure rates.

Pecht and Ramappan (1992) reviewed component and device (electronic system) field failure return data collected between 1970 and 1990.

TABLE D-6 Causes of Failure in Electronic Systems, in Percentage of Total Failures)

Category of Failure	Denson (1998)	Pecht and Nash (1994)
Parts	22	16
Design Related	9	21
Manufacturing Related	15	18
Externally Induced	12	-
Other (software, management, etc.)	22	17

Their analysis revealed that even in 1971, part failures only accounted for about 50 percent of the total failures in certain avionics systems. A similar analysis conducted in 1990 on avionics systems that had been deployed for 2–8 years revealed that the fraction of total failures caused by part failures was almost negligible (Pecht and Nash, 1994). Similar results were found by organizations such as Boeing and Westinghouse (between 1970 and 1990) including Bloomer (1989), Westinghouse Electric Corp. (1989), Taylor (1990), and Aerospace Industries Association (1991), and, more recently, by the U.S. Army Materiel Systems Analysis Activity (AMSAA) (Jais et al., 2013).

Lack of Consideration of Appropriate Environmental and Loading Conditions

The common stresses and conditions that cause failures in electronics include temperature cycling, mechanical bending, and vibration. MIL-HDBK-217 does not account for these different environmental and loading conditions. In other words, the handbook-based prediction methodology does not distinguish between the various kinds of stresses that a device or component might be subjected to in its field operating conditions. It also does not distinguish between different kinds of failures in the computation of failure rates. The handbook-based approaches assume that the device will continue to fail at the same rate (constant failure rate) under each of those operating conditions. Consequentially, the failure rate provided by the handbook-based prediction methodology is not useful in serving as indicator of the actual reliability of a device.

While the first edition of the MIL-HDBK-217 only featured a single-point constant failure rate, the second edition, MIL-HDBK-217B, featured a failure rate calculation that later became the standard for reliability prediction methodologies in the United States. In 1969, around the time that revision B of the handbook was being drafted, Codier (1969) wrote

> This traditional ritual is a flimsy house of cards which has almost no connection whatever with the 1969 realities of hardware development. The reason is, of course, that the failure rates are faulty. Frantic efforts are being made to bolster the theory by stress factors and environmental factors, but we cannot keep up. We are simply defining new constants whose values we cannot evaluate. The circulated draft of MIL-HDBK-217B proposes as many as five formal constants to use in determining the theoretical failure rate of a single part. In addition, is the following statement 'The many factors which constitute the base for any prediction can be listed briefly. . . . All these factors play an important part in the accuracy of forecast.' There follows a list of twenty-three factors.

Codier (1969) goes on to quote MIL-HDBK-217B: "In other words the forecast accuracy is based more on a prediction of program control effectiveness than it is on the inherent reliability of the design and its components." This can be illustrated by considering one of the equations from the handbook—in this particular example, we use the equation for the failure rate of bipolar junction transistors (BJTs):

$$\lambda_p = \lambda_b \pi_T \pi_R \pi_S \pi_Q \pi_E \; failures/10^6 \; hours, \tag{1}$$

where λ_b is the base failure rate, π_T is the temperature factor, π_R is the power rating factor, π_S is the voltage stress factor, π_Q is the quality factor, and π_E is the environment factor. Hence, it is assumed that environmental conditions can be accounted for by using multiplicative factors that scale the base failure rate linearly. The rationale behind this calculation is unclear; however, these types of calculations are available for almost all component types—both passive and active. Hence, the reason for Codier's skepticism is apparent—as there seems to be no scientific explanation as to why or how the factors are chosen, and why they have the values that they do. Furthermore, these factors do not specifically address the different failure modes and mechanisms that the device or component would be subject to under different loading or environment conditions.

The environment factor, π_E, in (1) does not directly account for specific stress conditions such as vibration, humidity, or bend. It represents generic conditions, such as ground benign, ground fixed, and ground mobile. These conditions refer to how controlled the environments generally are under different categories of military platforms. If we consider solder-level failures (in packages such as Ball Grid Arrays and others), there is little reported about failures of solders under high temperature "ageing" or "soak," even though these conditions could cause solders to be susceptible to failure under additional loads. However, solders are prone to failure in many ways, including in terms of thermal fatigue (temperature cycling) (see Ganesan and Pecht, 2006; Abtew and Selvaduray, 2000; Clech, 2004; Huang and Lee, 2008), vibrational loading (Zhou et al., 2010), mechanical bending (Pang and Che, 2007), and drop testing (Varghese and Dasgupta, 2007).

In addition, different solders exhibit different reliabilities and lifetimes. The time to failure also depends on the loading rates (thermal shock or thermal cycling; mechanical bend; and vibration or drop). None of these parameters is accounted for in the MIL-HDBK-217-based methodologies, as it does not factor in environmental and loading conditions. Not only would it be futile to try to document all the different types of solders and loading conditions, but also because of how quickly technology is advancing in these areas, it would be impossible to prepare an exhaustive list of combinations of loading conditions and solder types together with other parameters such as package type, board finish, etc. Hence, this is another area in which no solution to "repair" the approaches of the existing handbook is available.

Absence of Newer Technologies and Components

Since its introduction, MIL-HDBK-217 has had difficulty keeping up with the rapid advances being made in the electronics industry. The handbook underwent six revisions over the course of more than 30 years, with the last revision being MIL-HDBK-217F in 1995. Moore (1965) predicted that the complexity of circuits would increase in such a way that the number of transistors on integrated circuits would double nearly every 2 years. This means that the number of transistors on the generations of integrated circuits between 1965 and 1995 had increased by a factor of nearly 2^{15} (see Figure D-4).

The various handbook revisions were barely able to capture a small number of those generations of integrated circuits. Even with all its updates, the latest handbook revision, MIL-HDBK-217F, only features reliability prediction models for ceramic and plastic packages based on data from dual inline packages and pin grid arrays, which have rarely been used in new designs since 2003. Since the 1990s, the packaging and input/output (I/O) density of integrated circuits have advanced rapidly. Consequently, MIL-HDBK-217F, which does not differentiate between different package types or I/O densities, would not be applicable to any of the newer package types, including many of the new area array packages and surface mount packages. Hence, packages such as ball grid arrays (BGAs), quad-flat no-lead packages (QFN), package on package (PoP), and stacked die packages are not covered by MIL-HDBK-217F. Additionally, it would be a nearly impossible task to characterize the failure rates of all the different types of current generation packages, simply because of the sheer number of different packages.

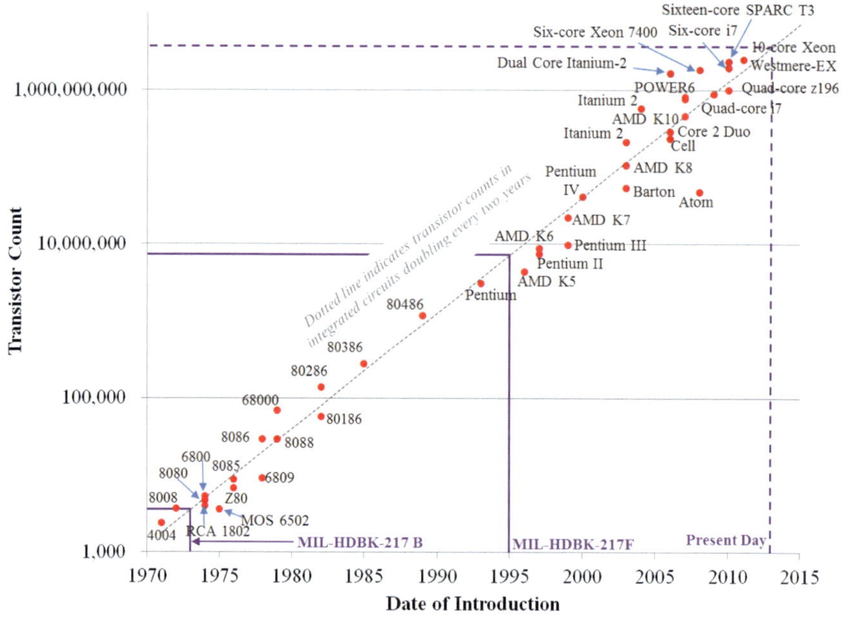

FIGURE D-4 Evolution of integrated circuits (Moore's Law).
SOURCE: Transistor Count and Moore's Law-2011 by Wgsimon (own work). Licensed under Creative Commons Attribution-Share Alike 3.0 via Wikimedia Commons. Available: http://commons.wikimedia.org/wiki/File:Transistor_Count_and_Moore%27s_Law_-_2011.svg#mediaviewer/File:Transistor_Count_and_Moore%27s_Law_-_2011.svg.

Moreover, even if it were possible to characterize the different package types and I/O densities, the life cycles (from design or conception to discontinuation) of modern electronic devices and components are much shorter than those of older components and systems. Some of the older package types, from the 1980s and 1990s, are still in use in legacy systems. Such components and systems had been made available for significantly longer than contemporary commercial electronic products, such as computers and cell phones, which typically have a life cycle of 2-5 years. These shorter life cycles pose a challenge to failure rate evaluations because of the short time frame available for the collection of failure data and the development of failure rate models. Hence, it would neither be pragmatic nor economical to invest in the development of such failure rate models.

The problem is not restricted only to active components: the discrete and passive component database of MIL-HDBK-217 is also outdated.

Discrete components, such as insulated gate bipolar transistors (IGBTs), surface mount tantalum capacitors, power sources such as Li-ion and NiMH batteries, supercapacitors, and niobium capacitors, all of which are used widely in the industry, are not, and probably will never be, covered by MIL-HDBK-217. Consequently, manufacturers and engineers are forced to identify the closest match in the handbook and then base their calculations on the guidelines prescribed for those older parts. This could potentially penalize a newer, more reliable component for the unreliability of its predecessors.

Historically, there have been precedents for such penalizations. For example, when the reliability, availability, and maintainability (RAM) model in MIL-HDBK-217B was extrapolated to include 64K RAM, the resulting mean time between failures calculated was a mere 13 seconds (Pecht and Nash, 1994). As a result of such incidents, a variety of notice changes to MIL-HDBK-217B appeared, and on April 9, 1979, MIL-HDBK-217C was published to "band aid" the problems. As a consequence of the rapidly advancing and ever-changing technology base, in the brief span of less than 3 years, MIL-HDBK-217C was updated to MIL-HDBK-217D in 1982, which in turn was updated to MIL-HDBK-217E in 1986. Hence, all the MIL-HDBK-217 revisions have had trouble keeping pace with the cutting edge of electronics packaging technology, and MIL-HDBK-217F is no exception. It would be easier to concede that the MIL-HDBK-217 methodology and approach are fundamentally and irreparably flawed than to update or replace it with something similar.

COMPARISONS OF HANDBOOK PREDICTIONS WITH FIELD DATA

Several case studies and experiments conducted have invalidated the predictions obtained from the handbook methodologies. These studies revealed that there were wild discrepancies between the predicted and actual MTBFs, often to several orders of magnitude. MTBFs and failures in time (FITs) are corollaries of the constant failure rate metric. For constant failure rates, the MTBF is simply the inverse of the constant failure rate, while an FIT rate is the number of failures in one billion (10^9) hours of device operation.

Some of the earliest studies that discovered the inconsistencies between test or field data and MIL-HDBK-217 predictions were published in the early 1960s (see Milligan, 1961; Pettinato and McLaughlin, 1961). The data from these tests indicated that the electronic systems of that time showed decreasing failure rates. However, as noted above, these findings were dismissed as anomalies. By the 1970s, there were more studies published—one of them identifying fluctuations in the field MTBF values when compared with the MIL-HDBK-217A-based predictions (Murata, 1975). In 1979, another study provided well-documented results that provided an indisputable challenge to constant failure rate models (Bezat and Montague, 1979).

Around this time, MIL-HDBK-217 was revised several times in an effort to keep up with the constantly evolving technology and to temporarily fix some of the models. However, despite the updates, there were more studies revealing disparities between test data and predicted MTBF values. One such study, conducted on Electric Countermeasures (ECM) radar systems, was carried out by Lynch and Phaller (1984). They pointed out that not only were the assumptions made in the calculation for reliability prediction at the part level flawed, but they also had a significant role in contributing to the disparity between predicted and observed MTBF values at the system level. Since then, several similar studies have been published, with the Annual Reliability and Maintainability Symposium including one such paper roughly every year (e.g., MacDiarmid, 1985; Webster, 1986; Branch, 1989; Leonard and Pecht, 1991; Miller and Moore, 1991; Rooney, 1994).

In the 1990s, even the proponents of the handbook-based techniques conceded that "MIL-HDBK-217 is not intended to predict field reliability and, in general, does not do a very good job of it in an absolute sense" (Morris, 1990). Companies such as General Motors stated that "GM concurs and will comply with the findings and policy revisions of Feb. 15, 1996 by the Assistant Secretary of the U.S. Army for Research, Development and Acquisition. . . . Therefore: MIL-HDBK 217, or a similar component reliability assessment method such as SAE PREL, shall not be used" (GM North America Operation, 1996). U.S. Army regulation 70-1 stated in a similar vein that "MIL–HDBK–217 or any of its derivatives are not to appear in a solicitation as it has been shown to be unreliable, and its use can lead to erroneous and misleading reliability predictions" (U.S. Department of the Army, 2011, pp. 15-16).

The following section reviews the results from some of the numerous case studies on both military and nonmilitary systems that have discredited the handbook-based techniques. The differences between the predictions of several similar handbook methodologies are also reviewed.

STUDIES ON COMMERCIAL ELECTRONIC PARTS AND ASSEMBLIES

There have been studies conducted on several different types of commercial electronic parts and assemblies, such as computer parts and memory (see Hergatt, 1991; Bowles, 1992; Wood and Elerath, 1994), industrial and automotive (Casanelli et a.l, 2005), avionics (Leonard and Pecht, 1989; Leonard and Pecht, 1991; Charpanel et al., 1998), and telecommunication equipment (Nilsson and Hallberg, 1997). Each of these studies reveals that there is a wide gap dividing the predicted MTBF values from the actual field or test MTBFs.

Table D-7 shows results from some publications on the disparity between the MTBFs for different devices. It can be seen that the ratios of measured MTBFs to predicted MTBFs varies from 0.54 to 12.20 for these systems.

Studies on computer systems, such as those by Wood and Elerath (1994) and Charpenel et al. (1998), seem to indicate that measured MTBFs are considerably lower than the predicted MTBF values. The results from their studies can be seen in Figure D-5. Similar results can also be seen in

TABLE D-7 Ratio of Measured MTBFs to Handbook-Based Predicted MTBFs for Various Electronic Devices

Product	Method	Measured MTBF (hours)	Predicted MTBF (hours)	Ratio
Audio Selector (Leonard and Pecht, 1991)		6,706	12,400	0.54
Bleed Air Control (Leonard and Pecht, 1991)		28,261	44,000	0.64
Storage (Hard) Disk (Wood and Elerath, 1994)	Bellcore* (*now Telcordia)			2.00
Processor Board (Wood and Elerath, 1994)	Bellcore*			2.50
Controller Board (Wood and Elerath, 1994)	Bellcore*			2.50
Power Supply (Wood and Elerath, 1994)	Bellcore*			3.50
PW2000 Engine Control (Leonard, 1991)		42,000	10,889	3.90
JT9D Engine Control (Leonard, 1991)		32,000	8,000	4.00
Memory Board (Wood and Elerath, 1994)	Bellcore*			5.00
Spoiler Control (Leonard and Pecht, 1991)		62,979	8,800	7.16
PC Server (Hergatt, 1991)	MIL-HDBK-217	15,600	2,070	7.50
Office Workstation (Hergatt, 1991)	MIL-HDBK-217	92,000	7,800	11.80
Avionics CPU Board	MIL-HDBK-217	243,902	20,450	11.90
Yaw Damper		55,993	4,600	12.20

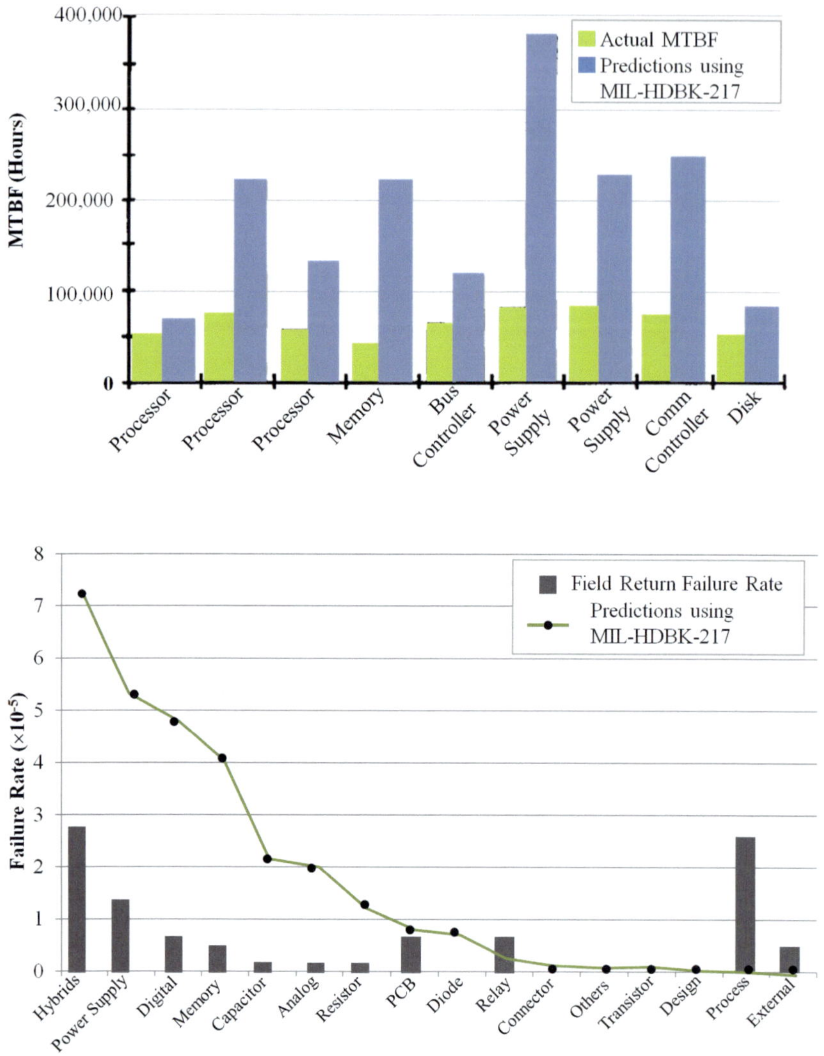

FIGURE D-5 Predicted MTBFs compared with field or measured MTBFs in Wood and Elerath (1994), top, and in Charpenel et al. (1998), bottom.
SOURCE: Top: Wood and Elerath (1994, p. 154). Reprinted with permission. Bottom: Charpenel et al. (1998). Reprinted with permission.

Hergatt (1991) regarding commercial computers, where the actual MTBFs of office workstations and PC servers were found to be three orders of magnitude greater than the predicted values. However, the results described in Table D-7 seem to indicate that handbook-based reliability prediction techniques can either arbitrarily underpredict or overpredict the true MTBF of a system in field conditions. Hence, handbook-based predictions are not always conservative. Even if the predictions were conservative, it would be impossible to determine the true margin between the predicted values and the actual values.

STUDIES ON MILITARY ELECTRONIC PARTS AND ASSEMBLIES

The errors in reliability prediction while using handbook-based methodologies are not restricted only to commercial electronics. A study conducted by the AMSAA (Jais et al., 2013) surveyed various agencies throughout the U.S. Department of Defense (DoD) by requesting reliability information on a variety of systems. The systems represented a variety of platforms, including communications devices, network command and control, ground systems, missile launchers, air command and control, aviation warning, and aviation training systems. The information requested included system-level predictions and demonstrated results (MTBFs from testing and fielding).

The results were filtered to only include estimates from predictions that were based purely on the methodologies prescribed in the MIL-HDBK-217 and its progeny. The ratio of predicted to demonstrated values ranged from 1.2:1 to 218:1: see Figure D-6. Jais et al. (2013, p. 4) stated that the "original contractor predictions for DoD systems [MTBFs] greatly exceed the demonstrated results." Statistical analysis of the data using Spearman's rank order correlation coefficient showed that MIL-HDBK-217-based predictions could not support comparisons between systems. The data and analysis demonstrated that the handbook predictions are not only inaccurate, but also could be harmful from an economic perspective if they were used as guidelines for sustainment, maintainability, and sparing calculations. The authors then considered why MIL-HDBK-217 is still being used for the DoD acquisition. They conclude that ". . . despite its shortcomings, system developers are familiar with MIL-HDBK-217 and its progeny. It allows them a 'one size fits all' tool that does not require additional analysis or engineering expertise. The lack of direction in contractual language leaves also government agencies open to its use" Jais et al. (2013, p. 5.)

Cushing et al. (1993) published data showing the discrepancies between the MIL-HDBK-217-based predictions of MTBFs for the Single Channel Ground Air Radio Set (SINCGARS) and the observed MTBFs during the 1987 nondevelopmental item candidate test. The data are shown in Table D-8. The errors in predictions were found to vary between –70 per-

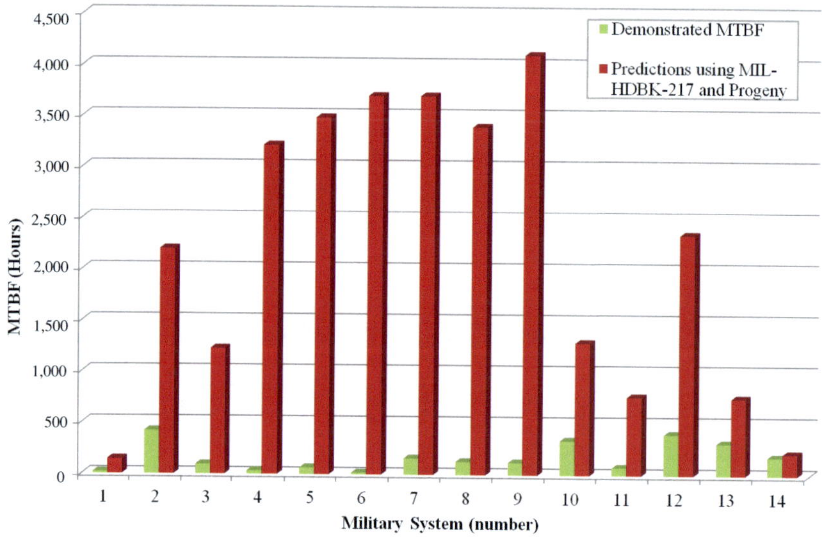

FIGURE D-6 Comparison of predicted and demonstrated MTBF values for U.S. defense systems.
SOURCE: Jais et al. (2013, p. 4). Reprinted with permission.

TABLE D-8 Results of the 1987 SINCGARS Nondevelopmental Item Candidate Test

Vendor	Predicted MTBF (hours)	Observed MTBF (hours)	Error (%)
A	811	98	728
B	1269	74	1615
C	1845	2174	–15
D	2000	624	221
E	2000	51	3822
F	2304	6903	–67
G	2450	472	419
H	2840	1160	145
I	3080	3612	–15

SOURCE: Cushing et al. (1993, p. 543). Reprinted with permission.

cent (underprediction) and 3,800 percent (overprediction). These data also highlight the issue with the ability of the handbook-based predictions to evaluate competing proposals. Hence, the column with the sorted predicted MTBF values does not provide an accurate ranking of the reliability of the vendors' systems, as vendor F's system was technically found to be more reliable than those of vendors G, H, and I and vendor E's system performed more poorly than those of vendors A, B, C, and D. These results again demonstrate how the handbook predictions can be misleading and inaccurate.

VARIATIONS IN PREDICTIONS BY DIFFERENT METHODOLOGIES

There are also variations in the predictions of MTBFs of systems obtained by different 217-type methodologies. Oh et al. (2013) found that when predicting the reliability of cooling fans and their control systems using MIL-HDBK-217 and Telcordia SR-332, very different results were obtained. While the MTBF predictions based on MIL-HDBK-217 showed an error of 469–663 percent, the SR-332 prediction had an error of between 70 and 300 percent. Spencer (1986) compared the FIT rates of NMOS SRAM modules evaluated using MIL-HDBK-217, Bellcore (Telcordia), British Telecom HRD, and CNET based methodologies. He found that the failure rate increased with an increase in complexity for all the prediction methodologies.

Jones and Hayes (1999) performed a more comprehensive study comparing predictions from various handbook methodologies with actual field data on commercial electronic circuit boards. They found not only a difference between the predictions and MTBFs evaluated using the field data, but also significant differences between the various predictions themselves. Figure D-7 provides some details.

RIAC 217PLUS AND MIL-HDBK-217G

The Handbook of 217Plus Reliability Prediction Models, commonly referred to as 217Plus, was developed by the Reliability Information Analysis Center (RIAC), an independent private enterprise. 217Plus was written to document the models and equations used in PRISM, a software tool used for system reliability assessment, described above. 217Plus was also published as an alternative to MIL-HDBK-217F. 217Plus prediction model doubles the number of part-type failure rate models from PRISM, and it also contains six new constant failure rate models not available in PRISM. The Handbook of 217Plus was released on May 26, 2006.

217Plus contains reliability prediction models for both the component and system levels. The component models are determined first to estimate the failure rate of each component and then summed to estimate the sys-

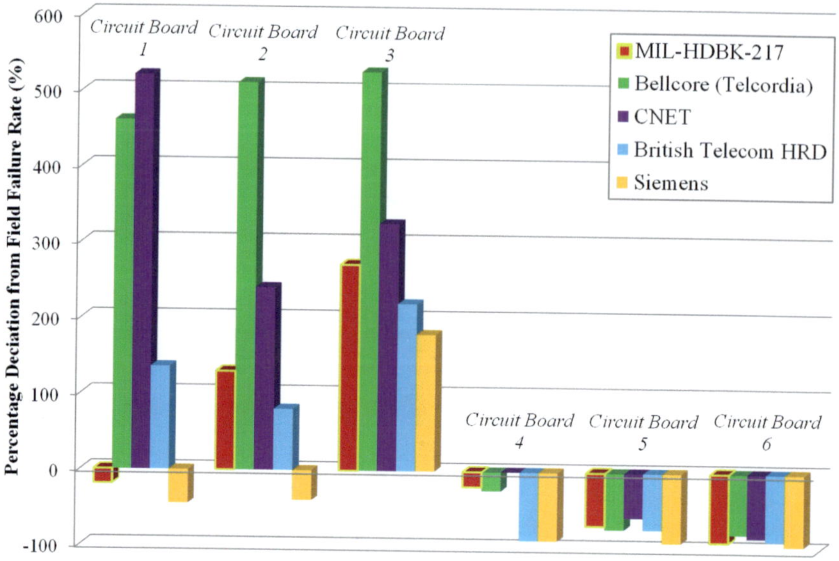

FIGURE D-7 Comparison of various handbook methodologies.
SOURCE: Adapted from Jones and Hayes (1999, p. 9). Reprinted with permission.

tem failure rate. This estimate of system reliability is further modified by the application of "system-level" factors (called process grade factors) that account for noncomponent impacts of overall system reliability. "The goal of a model is to estimate the 'rate of occurrence of failure' and accelerants of a component's primary failure mechanisms within an acceptable degree of accuracy" (Reliability Information Analysis Center, 2006, p. 2). The models account for environmental factors and operational profile factors so that various tradeoff analyses can be performed. A 217Plus prediction can be performed using both a predecessor system and a new system. A predecessor is a product with similar technology and design and manufacturing processes. If the item under analysis is an evolution of a predecessor item, then the field experience of the predecessor item can be leveraged and modified to account for the differences between the new item and the predecessor item. The 217Plus methodology also accommodates the incorporation of test/field reliability experience into the analytical prediction of new systems.

Even if RIAC 217Plus overcomes some of the major shortcomings of MIL-HDBK-217, it is still built on the foundations of the handbook. As it stands now, it is unclear how 217Plus accounts for all possible equipment failure modes by focusing on component models and process grade factors. Although additive models help curb the "blow up" of MTTF values calculated at the high and low extremes, it still remains to be proven how this would provide more accurate estimates of failure rates and lifetimes. The implementation of such modifications as the additive models still does not account for common mode failures, which account for a significant portion of system-level failures. It is also unclear if 217Plus can account for the dependencies between the various components and their operations under different environmental conditions at the system level. Combining the failure rates estimated theoretically (by considering the system to be a superset of components) with results from Bayesian analysis and estimation of the failure rates using field and test data would account for some of these common mode failures. However, the validity and applicability of these failure rates depends on the assumptions made, specifically regarding the likelihood functions involved in the Bayesian analysis. The final calculations and estimates of lifetimes and failure rates also depend on how the results from the Bayesian analysis are merged with other analytical predictions. It is still not be possible to condense the results from the Bayesian analysis to give a point estimate of either the lifetimes or failure rates, because this point estimate would suffer from the same shortcomings that the constant failure rate estimates have, in that it would not be able to account for variations in field conditions.

Although the 217Plus methodology was developed by RIAC, the inclusion of "217" term in the title of the RIAC handbook seems to imply that it is officially endorsed by DoD as a successor to MIL-HDBK-217F. But 217Plus is far from being a successor to MIL-HDBK-217F. The 217Plus would have been capable of serving as an interim alternative to MIL-HDBK-217F as it is being phased out, if 217Plus were able to address system-level reliability and reliability growth. Its predictions would also need to take into consideration the physics of failure both at the component and system levels. However, because 217Plus is not capable of addressing system-level reliability or physics-of-failure issues adequately, it cannot really serve either as the successor or an interim alternative to MIL-HDBK-217F. The remedy to the damage caused by adopting MIL-HDBK-217 is not to substitute it with a "lesser evil," but rather to do away with all such techniques.

The Naval Surface Warfare Center (NSWC) Crane Division announced its intent to form and chair a working group to revise MIL-HDBK-217F to Revision G by updating the list of components and technologies. But, as Jais et al., (2013, p. 5) have stated, "Simply updating MIL-HDBK-217 based upon current technology does not alleviate the underlying fundamental technical limitations addressed in the earlier sections. Predictions should provide design information on failure modes and mechanisms that can be used to mitigate the risk of failure by implementing design changes." The NSWC Crane Division recognizes that MIL-HDBK-217 is known and accepted worldwide and is used by commercial companies, the defense industry, and government organizations. Instead of educating users about the shortcomings of MIL-HDBK-217 techniques and assumptions, it prescribes standardization of the use of the 217 reliability prediction tool. The NSWC also stated that the end users of the 217 methodology prefer the relative simplicity of the prediction method. This approach demonstrates how the usage of the MIL-HDBK-217 methodologies has desensitized the reliability engineering community and the electronics industry as a whole. A new paradigm shift is needed, and the emphasis should not be on the relative simplicity of a reliability prediction methodology, but rather on the scientific merit and accuracy of the techniques.

SUMMARY AND CONCLUSIONS

MIL-HDBK-217 is a reliability prediction methodology for electronic components and devices that is known to be fundamentally flawed in many ways. The problems with the use of MIL-HDBK-217 can broadly be classified into two categories. The first category of problems arise from the fact that the MIL-HDBK-217 concept was developed, formalized, and institutionalized before the knowledge of electronics and degradation had become mature. Thus, the first version of the handbook featured a simplistic single point constant failure obtained by fitting a straight line through field failure data of vacuum tubes. Subsequent revisions of the handbook were influenced by the exponential distributions and the associated constant failure rates that were being used in actuarial studies. This constant failure rate then became the premise of MIL-HDBK-217. Because the methodology was built around this premise, the reliability prediction techniques excluded any consideration of the root causes of failures and the physics underlying the failure mechanisms and focused instead only on the linear regression analysis of the failure data. As a result, the handbook did not have guidelines or calculations to account for different types of environmental and operational loading conditions, and the predicted failure rates were assumed to be the same constant value in all conditions

The second category of problems arise from the fact that for a technique such as the MIL-HDBK-217 methodology, it becomes increasingly difficult to keep up to date with the fast pace of the development of electronics and changes in the supply-chain balance. With constant advances in packaging and interconnected technology and materials, both active and passive electronic components are evolving rapidly. Components such as insulated gate bipolar transistors and niobium capacitors, which are now used extensively in electronic systems, were not included in the last MIL-HDBK-217 revision. Furthermore, the life cycle of components in the 2010s is vastly shorter than the life cycle of components in the 1960s. With the average life cycle of commercial electronic systems, such as smartphones and laptops, being about 2 years, it becomes very difficult to acquire field failure data and build failure models for every type of component used in electronic systems.

The progeny of MIL-HDBK-217 were all developed and promoted on the assumption that the only problem with the original methodology was that it was not up to date or that it did not meet the needs of a specific market sector. The progeny include such methodologies as SAE Reliability Prediction, Bellcore/Telcordia, PRISM, and RIAC 217Plus. While each of these methodologies might have application-specific test conditions or data for newer components that were excluded from MIL-HDBK-217, they all still use the constant failure rate assumption in some capacity. Hence, each of these "new" approaches continued to ignore the fundamental scientific principles that govern degradation and failure mechanics of electronic devices. These progeny, like their predecessor, failed to acknowledge that the degradation and failure of a component cannot be condensed into a single unique "constant failure rate" metric.

The inability of the MIL-HDBK-217 methodology and its progeny to predict failure rates has been demonstrated by numerous studies. In each of these studies the failure rate predictions end up either grossly overpredicting or underpredicting the actual failure rates. Therefore, we conclude that the MIL-HDBK-217 approach provides the user with values that are inaccurate and misleading. However, because adhering to MIL-HDBK-217 for failure rate calculations has often been a contractual requirement, and because this methodology was also adapted and used by the telecommunication and automobile industries, the practice of using the predictions based on these techniques trickled down the supply chain, thus pervading the electronics industry. This practice culminated in the proliferation of MIL-HDBK-217. As a consequence, the use of constant failure rate models is widespread even today, despite the fact that the handbook has not been updated in nearly 20 years. A NSWC CRANE survey in 2008 (Gullo, 2008) reported that 80 percent of its respondents still used MIL-HDBK-217.

The adoption and adaptation of constant failure rate models to evaluate the reliability of electronic systems was probably never a good idea. This practice has fundamentally affected how reliability prediction is perceived, both with regard to commercial and military electronics. The continued use of MIL-HDBK-217 or one of its adaptations can be destructive because it promotes poor engineering practices while also harming the growth of reliability of electronic products. Furthermore, MIL-HDBK-217 cannot be improved or fixed because the underlying assumptions governing the methodology are fallacious. It must be accepted that a reliability prediction methodology that is based on predicting the use conditions is not feasible. Moving forward, the solution is to do away with MIL-HDBK-217 by canceling and no longer recognizing it. DoD should strive for a policy whereby every major subsystem and critical component used in a defense system have physics-of-failure models for component reliability that have been validated by the manufacturer.

REFERENCES

Abtew, M., and Selvaduray, G. (2000). Lead-free solders in microelectronics. *Materials Science and Engineering Reports, 27*(5-6), 95-141.

Aerospace Industries Association. (1991). *Ultra Reliable Electronic Systems—Failure Data Analysis by Committee of AIA Member Companies.* Arlington, VA: Author.

Al-Sheikhly, M., and Christou, A. (1994). How radiation affects polymeric materials. *IEEE Transactions on Reliability, 43,* 551-556.

Asher, H., and Feingold, F.H. (1984). *Repairable Systems Reliability: Modeling, Inference, Misconceptions and Their Causes, Lecture Notes in Statistics, Volume 7.* New York: Marcel Dekker.

Bar-Cohen, A. (1988). Reliability physics vs. reliability prediction. *IEEE Transactions on Reliability, 37,* 452.

Bezat, A.G., and Montague, L.L. (1979). The effects of endless burn-in on reliability growth projections. Prepared for the Annual Reliability and Maintainability Symposium, January 23-25, Washington, DC. In *Proceedings of the 1979 Reliability and Maintainability Symposium* (pp. 392-397). New York: IEEE.

Bhagat, W.W. (1989). R&M through avionics/electronics integrity program. Prepared for the Annual Reliability and Maintainability Symposium, January 24-26, Atlanta GA. In *Proceedings of the 1989 Reliability and Maintainability Symposium* (pp. 216-220). New York: IEEE. Available: http://ieeexplore.ieee.org/stamp/stamp.jsp?arnumber=49604 [December 2014].

Blanks, H.S. (1980). The temperature dependence of component failure rate. *Microelectronics and Reliability, 20,* 297-307.

Bloomer, C. (1989). Failure mechanisms in through hole packages. *Electronic Materials Handbook, 1,* 976.

Bowles, J.B. (1992). A survey of reliability-prediction procedures for microelectronic devices. *IEEE Transactions on Reliability, 41*(1), 2-12.

Bowles, J.B. (2002). Commentary—caution: Constant failure-rate models may be hazardous to your design. *IEEE Transactions on Reliability, 51*(3), 375-377.

Caruso, H. (1996). An overview of environmental reliability testing. Prepared for the Annual Reliability and Maintainability Symposium, January 22-25. In *Proceedings of the 1996 Reliability and Maintainability Symposium* (pp. 102-109). New York: IEEE. Available: http://ieeexplore.ieee.org/stamp/stamp.jsp?arnumber=500649 [December 2014].

Cassanelli, G., Mura, G., Cesaretti, F., Vanzi, M., and Fantini, F. (2005). Reliability predictions in electronic industry applications. *Microelectronics Reliability, 45,* 1321-1326.

Charpenel, P., Cavernes, P., Casanovas, V., Borowski, J., and Chopin, J.M. (1998). Comparison between field reliability and new prediction methodology on avionics embedded electronics. *Microelectronics Reliability, 38,* 1171-1175.

Clech, J. (2004). Lead-free and mixed assembly solder joint reliability trends. *Proceedings of the 2004 IPC/SMEMA Council APEX Conference,* Anaheim, CA. Available: http://www.jpclech.com/Clech_APEX2004_Paper.pdf [October 2014].

Codier, E.O. (1969). Reliability prediction—Help or hoax? *Proceedings of the 1969 Annual Symposium on Reliability.* IEEE Catalog No 69C8-R, pp. 383-390. New York: IEEE.

Coleman, L.A. (1992). Army to abandon MIL-HDBK-217. *Military and Aerospace Electronics, 3*, 1-6.

Cushing, M.J., Mortin, D.E., Stadterman, T.J., and Malhotra, A. (1993). Comparison of electronics reliability assessment approaches. *IEEE Transactions on Reliability, 42*(4), 542-546.

Cushing, M.J., Krolewski, J.G., Stadterman, J.T., and Hum, B.T. (1996). U.S. Army reliability standardization improvement policy and its impact. *IEEE Transactions on Components, 19*(2), 277-278.

Dasgupta, A. (1993). Failure-mechanism models for cyclic fatigue. *IEEE Transactions on Reliability, 42*, 548-555.

Dasgupta, A., and Haslach, H.W.J. (1993). Mechanical design failure models for buckling. *IEEE Transactions on Reliability, 42*, 9-16.

Dasgupta, A., and Hu, J.M. (1992a). Failure-mechanism models for excessive elastic deformation. *IEEE Transactions on Reliability, 41*, 149-154.

Dasgupta, A., and Hu, J.M. (1992b). Failure-mechanism models for plastic deformations. *IEEE Transactions on Reliability, 41*, 168-174.

Dasgupta, A., and Pecht, M. (1991). Material failure-mechanisms and damage models. *IEEE Transactions on Reliability, 40*, 531-536.

Denson, W. (1998). The history of reliability prediction. *IEEE Transactions on Reliability, 47*(3), SP3211-SP3218.

Denson, W., and Brusius, P. (1989). *VHSIC and VHSIC-like reliability prediction modeling*. RADC-TR-89-177, Final Technical Report. Rome, NY: IIT Research Institute.

Diaz, C., Kang, S.M., and Duvvury, C. (1995). Tutorial: Electrical overstress and electrostatic discharge. *IEEE Transactions on Reliability, 44*, 2-5.

Engel, P. (1993). Failure models for mechanical wear modes and mechanisms. *IEEE Transactions on Reliability, 42*, 262-267.

Epstein, B., and Sobel, M. (1954). Some theorems relevant to life testing from an exponential distribution. *The Annals of Mathematical Statistics, 25*, 373-381.

FIDES Group. (2009). *FIDES Guide 2009*. Paris, France: Union Technique De L'Electricite.

Ganesan, S., and Pecht, M. (2006). *Lead-Free Electronics*. Hoboken, NJ: John Wiley & Sons.

GM North America Operation. (1996). *Technical Specification Number: 10288874*. Detroit, MI: General Motors Company.

Gu, J., and Pecht, M. (2007). New methods to predict reliability of electronics. *Proceedings of the Seventh International Conference on Reliability and Maintainability* (pp. 440-451. Beijing: China Astronautic Publishing House.

Gullo, L. (2008). The revitalization of MIL-HDBK-217. *IEEE Transactions on Reliability, 58*(2), 210-261. Available: http://rs.ieee.org/images/files/Publications/2008/2008-10.pdf [January 2015].

Hallberg, Ö. (1994). Hardware reliability assurance and field experience in a telecom environment. *Quality and Reliability Engineering International, 10*(3), 195-200.

Hallberg, Ö., and Löfberg, J. (1999). A Time Dependent Field Return Model for Telecommunication Hardware, in *Advances in Electronic Packaging 1999: Proceedings of the Pacific Rim/ASME International Intersociety Electronic and Photonic Packaging Conference (InterPACK '99)*, New York.

Halverson, M., and Ozdes, D. (1992). What happened to the system perspective in reliability. *Quality and Reliability Engineering International, 8*(5), 391-412.

Hergatt, N.K. (1991). Improved reliability predictions for commercial computers. *Proceedings of the 1991 Annual Reliability and Maintainability Symposium*. Available: http://ieeexplore.ieee.org/stamp/stamp.jsp?arnumber=154461 [December 2014].

Huang, M., and Lee, C. (2008). Board level reliability of lead-free designs of BGAs, CSPs, QFPs and TSOPs. *Soldering and Surface Mount Technology, 20*(3), 18-25.

IEEE Standards Association. (2010). *Guide for Developing and Assessing Reliability Predictions Based on IEEE Standard 1413.* Piscataway, NJ: Author.

Jais, C., Werner, B., and Das, D. (2013). Reliability predictions: Continued reliance on a misleading approach. Prepared for the Annual Reliability and Maintainability Symposium, January 28-31, Orlando, FL. In *Proceedings of the 2013 Reliability and Maintainability Symposium* (pp. 1-6). New York: IEEE. Available: http://ieeexplore.ieee.org/stamp/stamp.jsp?tp=&arnumber=6517751 [December 2014].

Jensen, F. (1995). Reliability: The next generation. *Microelectronics Reliability, 35*(9-10), 1363-1375.

Jones, J., and Hayes, J. (1999). A comparison of electronic reliability prediction models. *IEEE Transactions on Reliability, 48*(2), 127-134.

Kelly, M., Boulton, W., Kukowski, J., Meieran, E., Pecht, M., Peeples, J., and Tummala, R. (1995). *Electronic Manufacturing and Packaging in Japan.* Baltimore, MD: Japanese Technology Evaluation Center and International Technology Research Institute.

Knowles, I. (1993). Is it time for a new approach? *IEEE Transactions on Reliability, 42,* 3.

Kopanski, J.J., Blackburn, D.L., Harman, G.G., and Berning, D.W. (1991). *Assessment of Reliability Concerns for Wide-Temperature Operation of Semiconductor Devices and Circuits.* Gaithersburg, MD: National Institute of Standards and Technology. Available: http://www.nist.gov/manuscript-publication-search.cfm?pub_id=4347 [September 2014].

Lall, P., and Pecht, M. (1993). An integrated physics of failure approach to reliability assessment. *Proceedings of the 1993 ASME International Electronics Packaging Conference, 4*(1).

Lall, P., Pecht, M., and Hakim, E. (1997). *Influence of Temperature on Microelectronics and System Reliability.* Boca Raton, FL: CRC Press.

Leonard, C.T. (1987). Passive cooling for avionics can improve airplane efficiency and reliability. *Proceedings of the IEEE 1989 National Aerospace and Electronics Conference.* New York: IEEE.

Leonard, C.T. (1988). On US MIL-HDBK-217. *IEEE Transactions on Reliability, 37*(1988), 450-451.

Leonard, C.T. (1991). Mechanical engineering issues and electronic equipment reliability: Incurred costs without compensating benefits. *Journal of Electronic Packaging, 113,* 1-7.

Leonard, C.T., and Pecht, M. (1989). Failure prediction methodology calculations can mislead: Use them wisely, not blindly. *Proceedings of the IEEE 1989 National Aerospace and Electronics Conference,* pp. 1887-1892.

Leonard, C.T., and Pecht, M. (1991). Improved techniques for cost effective electronics. *Proceedings of the 1991 Annual Reliability and Maintainability Symposium.*

Li, J., and Dasgupta, A. (1993). Failure-mechanism models for creep and creep rupture. *IEEE Transactions on Reliability, 42,* 339-353.

Li, J., and Dasgupta, A. (1994). Failure-mechanism models for material aging due to interdiffusion. *IEEE Transactions on Reliability, 43,* 2-10.

Li, X., Qin, J., and Bernstein, J.B. (2008). Compact modeling of MOSFET wearout mechanisms for circuit-reliability simulation. *IEEE Transaction on Device Materials and Reliability, 8*(1), 98-121.

Lynch, J.B., and Phaller, L.J. (1984). Predicted vs test MTBFs... Why the disparity? *Proceedings of the 1984 Annual Reliability and Maintainability Symposium,* pp. 117-122.

MacDiarmid, P.R. (1985). Relating factory and field reliability and maintainability measures. *Proceedings of the 1985 Annual Reliability and Maintainability Symposium,* p. 576.

McLinn, J.A. (1989). Is Failure Rate Constant for a Complex System? *Proceedings of the 1989 Annual Quality Congress,* pp. 723-728.

McLinn, J.A. (1990). Constant failure rate—A paradigm in transition. *Quality and Reliability Engineering International, 6,* 237-241.

Miller, P.E., and Moore, R.I. (1991). Field reliability vs. predicted reliability: An analysis of root causes for the difference. *Proceedings of the 1991 Annual Reliability and Maintainability Symposium,* pp. 405-410.

Milligan, G.V. (1961). Semiconductor failures Vs. removals. In J.E. Shwop, and H.J. Sullivan, *Semiconductor Reliability.* Elizabeth, NJ: Engineering.

Mishra, S., Ganesan, S., Pecht, M., and Xie, J. (2004). Life consumption monitoring for electronics prognostics. *Proceedings of the 2004 IEEE Aerospace Conference,* pp. 3455-3467.

Montgomery, D.C., and Runger, G.C. (1994). *Applied Statistics and Probability for Engineers.* New York: John Wiley & Sons.

Moore, G.E. (1965). Cramming more components onto integrated circuit. *Electronics Magazine, 38*(8), 4-7.

Morris, S. (1990). MIL-HDBK-217 use and application. *Reliability Review, 10,* 10-13.

Murata, T. (1975). Reliability case history of an airborne air data computer. *IEEE Transactions on Reliability, R-24*(2), 98-102.

Nash, F.R. (1993). *Estimating Device Reliability: Assessment of Credibility.* Boston, MA: Kluwer Academic.

Nilsson, M., and Hallberg, Ö. (1997). A new reliability prediction model for telecommunication hardware. *Microelectronics Reliability, 37*(10-11), 1429-1432.

O'Connor, P.D.T. (1985a). Reliability: Measurement or management. *Reliability Engineering, 10*(3), 129-140.

O'Connor, P.D.T. (1985b). Reliability prediction for microelectronic systems. *Reliability Engineering, 10*(3), 129-140.

O'Connor, P.D.T. (1988). Undue faith in US MIL-HDBK-217 for Reliability Prediction. *IEEE Transactions on Reliability, 37,* 468-469.

O'Connor, P.D.T. (1990). Reliability prediction: Help or hoax. *Solid State Technology, 33,* 59-61.

O'Connor, P.D.T. (1991). Statistics in quality and reliability—Lessons from the past, and future opportunities. *Reliability Engineering and System Safety, 34*(1), 23-33.

O'Connor, P.D.T., and Harris, L.N. (1986). Reliability prediction: A state-of-the-art review. *IEE Proceedings A (Physical Science, Measurement and Instrumentation, Management and Education, Reviews), 133*(4), 202-216.

Oh, H., Azarian, M.H., Das, D., and Pecht, M. (2013). A critique of the IPC-9591 standard: Performance parameters for air moving devices. *IEEE Transactions on Device and Materials Reliability, 13*(1), 146-155.

Pang, J., and Che, F.-X. (2007). Isothermal cyclic bend fatigue test method for lead-free solder joints. *Journal of Electronic Packaging, 129*(4), 496-503.

Patil, N., Celaya, J., Das, D., Goebel, K., and Pecht, M. (2009). Precursor parameter identification for insulated gate bipolar transistor (IGBT) prognostics. *IEEE Transactions on Reliability, 58*(2), 271-276.

Pease, R. (1991). What's all this MIL-HDBK-217 stuff anyhow? *Electronic Design, 1991,* 82-84.

Pecht, M., and Gu, J. (2009). Physics-of-failure-based prognostics for electronic products *Transactions of the Institute of Measurement and Control,* 31(3-4), 309-322. Available: http://tim.sagepub.com/content/31/3-4/309.full.pdf+html [December 2014].

Pecht M.G., and Nash, F.R. (1994). Predicting the reliability of electronic equipment. *Predicting the IEEE, 82*(7), 992-1004.

Pecht, M.G., and Ramappan, V. (1992). Are components still the major problem: A review of electronic system and device field failure returns. *IEEE Transactions on Components, Hybrids and Manufacturing Technology, 15*(6), 1160-1164.

Pecht, M.G., Lall, P., and Hakim, E. (1992). Temperature dependence of integrated circuit failure mechanisms. *Quality and Reliability Engineering International, 8*(3), 167-176.

Peck, D.S. (1971). The analysis of data from accelerated stress tests. *Proceedings of the 1971 Annual Reliability Physics Symposia,* pp. 69-78.

Pettinato, A.D., and McLaughlin, R.L. (1961). Accelerated reliability testing. *Proceedings of the 1961 National Symposium of Reliability and Quality,* pp. 241-251.

Ramakrishnan, A., and Pecht, M. (2003). A life consumption monitoring methodology for electronic systems. *IEEE Transactions on Components and Packaging Technology, 26*(3), 625-634.

Reliability Assessment Center. (2001). *PRISM—Version 1.3, System Reliability Assessmend Software.* Rome, NY: Reliability Assessment Center.

Reliability Information Analysis Center. (2006). *Handbook of 217Plus Reliability Models.* Utica, New York: Reliability Information Analysis Center.

Rooney, J.P. (1994). Customer satisfaction. *Proceedings of the 1994 Annual Reliability and Maintainability Symposium,* pp. 376-381.

Rudra, B., and Jennings, D. (1994). Tutorial: Failure-mechanism models for conductive-filament formation. *IEEE Transactions on Reliability, 43,* 354-360.

Ryerson, C.M. (1982). The reliability bathtub curve is vigorously alive. *Proceedings of the Annual Reliability and Maintainability Symposium,* p. 187.

Saleh J.H., and Marais, K. (2006). Highlights from the early (and pre-) history of reliability engineering. *Reliability Engineering and System Safety, 91,* 249-256.

Shetty, V., Das, D., Pecht, M., Hiemstra, D., and Martin, S. (2002). Remaining life assessment of shuttle remote manipulator system end effector. *Proceedings of 22nd Space Simulation Conference,* Ellicott City, MD. Available: http://www.prognostics.umd.edu/calcepapers/02_V.Shetty_remaingLifeAssesShuttleRemotemanipulatorSystem_22ndSpaceSimulationConf.pdf [October 2014].

Spencer, J. (1986). The highs and lows of reliability prediction. *Proceedings of the 1986 Annual Reliability and Maintainability Symposium,* pp. 152-162.

Taylor, D. (1990). Temperature dependence of microelectronic devices failures. *Quality and Reliability Engineering International, 6*(4), 275.

Telcordia Technologies. (2001). *Special Report SR-332: Reliability Prediction Procedure for Electronic Equipment, Issue 1.* Piscataway, NJ: Telcordia Customer Service.

U.S. Department of Defense. (1991). *MIL-HDBK-217. Military Handbook: Reliability Prediction of Electronic Equipment.* Washington, DC: Author. Available: http://www.sre.org/pubs/Mil-Hdbk-217F.pdf [August 2014].

U.S. Department of the Army. (2011). *Army Regulation 70-1: Army Acquisition Policy.* Washington DC: U.S. Department of the Army.

Varghese, J., and Dasgupta, A. (2007). Test methodology for durability estimation of surface mount interconnects under drop testing conditions. *Microelectronics Reliability, 47*(1), 93-107.

Watson, G.F. (1992). MIL reliability: A new approach. *IEEE Spectrum, 29*, 46-49.

Webster, L.R. (1986). Field vs. predicted for commercial SatCom terminals. *Proceedings of the 1986 Annual Reliability and Maintainability Symposium,* pp. 89-91.

Westinghouse Electric Corporation. (1989). *Summary Chart of 1984/1987 Failure Analysis Memos.* Cranberry Township, PA: Westinghouse Electric Corporation.

Wong, K.D., and Lindstrom, D.L. (1989). Off the bathtub onto the roller coaster curve. *Proceedings of the Annual Reliability and Maintainability Symposium,* January 26-28, Los Angeles, CA.

Wong, K.L. (1981). Unified field (failure) theory: Demise of the bathtub curve. *Proceedings of the 1981 Annual Reliability and Maintainability Symposium,* pp. 402-403.

Wong, K.L. (1989). The bathtub curve and flat earth society. *IEEE Transactions on Reliability, 38,* 403-404.

Wong, K.L. (1990). What is wrong with the existing reliability prediction methods? *Quality and Reliability Engineering International,* 6(4), 251-257.

Wong, K.L. (1993). A change in direction for reliability engineering is long overdue. *IEEE Transactions on Reliability, 42,* 261.

Wood, A.P., and Elerath, J.G. (1994). A comparison of predicted MTBFs to field and test data. *Proceedings of the 1994 Annual Reliability and Maintainability Symposium,* pp. 153-156.

Young, D., and Christou, A. (1994). Failure mechanism models for electromigration. *IEEE Transactions on Reliability, 43,* 186-192.

Zhou, Y., Al-Bassyiouni, M., and Dasgupta, A. (2010). Vibration durability assessment of Sn3.0Ag0.5Cu and Sn37Pb solders under harmonic excitation. *IEEE Transactions on Components and Packaging Technologies, 33*(2), 319-328.

Appendix E

Biographical Sketches of Panel Members and Staff

ARTHUR FRIES (*Chair*) is a research staff member and project leader at the Institute for Defense Analyses. He has focused on applying statistical methods to issues in various national defense and security sectors—test and evaluation within the U.S. Departments of Defense and Homeland Security, counter-narcotics, counter-terrorism, and risk assessment. He was chair of the American Statistical Association (ASA) Committee on National and International Security, Chair of the ASA Committee on Statisticians in Defense and National Security, and a founding member of the ASA Section on Statistics in Defense and National Security. He is a fellow of the ASA and recipient of the U.S. Army Wilks Award. He holds an M.A. in mathematics and a Ph.D. in statistics, both from the University of Wisconsin–Madison.

W. PETER CHERRY is chief analyst at Science Applications International Corporation. His work has focused on the development and application of operations research in the national security domain, primarily in the field of land combat. He contributed to the development and fielding of most of the major systems currently employed by the Army, ranging from the Patriot missile to the Apache helicopter. He was a member of the Army Science Board and served as chair of the Military Applications Society of the Operations Research Society of America. He is a co-awardee of the Rist Prize of the Military Operations Research Society and an awardee of the Steinhardt Prize of the Military Applications Society of the Institute for Operations Research and the Management Sciences. He is a member of the

National Academy of Engineering. He holds an M.S. and Ph.D. in industrial and operations engineering from the University of Michigan.

MICHAEL L. COHEN (*Study Director*) is a senior program officer for the Committee on National Statistics, where he directs studies involving statistical methodology, in particular on defense system testing and decennial census methodology. Formerly, he was a mathematical statistician at the Energy Information Administration, an assistant professor in the School of Public Affairs at the University of Maryland, and a visiting lecturer at Princeton University. He is a fellow of the American Statistical Association. He holds a B.S. in mathematics from the University of Michigan and an M.S. and a Ph.D. in statistics from Stanford University.

ROBERT G. EASTERLING was a senior statistical scientist at the Sandia National Laboratories, where he spent most of his career investigating the applications of statistics to various engineering issues. One of his key research interests has been reliability evaluation. He is a fellow of the American Statistical Association, a former editor of the applied statistics journal *Technometrics*, and a recipient of the American Society for Quality's Brumbaugh Award. He holds a Ph.D. in statistics from Oklahoma State University.

ELSAYED A. ELSAYED is a distinguished professor of the Department of Industrial and Systems Engineering and a fellow in the Rutgers Business, Engineering, Science and Technology Institute, both at Rutgers University. He is also director of the Industry/University Cooperative Research Center for Quality and Reliability Engineering, under the aegis of the National Science Foundation. His research interests are in the areas of quality and reliability engineering and production planning and control. He is a fellow of the Institute of Industrial Engineers. He holds a Ph.D. from the University of Windsor (Canada).

APARNA V. HUZURBAZAR is a research scientist in the Statistical Sciences Group at Los Alamos National Laboratory. At Los Alamos, she also serves as project lead for the Systems Major Technical Elements Enhanced Surveillance Campaign, which provides statistical and analytical support, such as system modeling, age-aware models, tracking and trending data, and uncertainty quantification. She has published extensively in reliability methodology and applications, flowgraph models, Bayesian statistics, and quality control and industrial statistics. She is a fellow of the American Statistical Association and an elected member of the International Statistical Institute. She holds a Ph.D. in statistics from Colorado State University.

PATRICIA A. JACOBS is a distinguished professor in the Department of Operations Research at the Naval Postgraduate School. For most of her career, she has focused on defense issues involving statistics and operations research, including reliability modeling. She is a fellow of the American Statistical Association and the Royal Statistical Society and is an elected member of the International Statistical Institute. She holds an M.S. in industrial engineering and management sciences and a Ph.D. in applied mathematics, both from Northwestern University.

WILLIAM Q. MEEKER, JR. is a distinguished professor of liberal arts and sciences and professor in the Department of Statistics at Iowa State University. His major research has been on statistical methods for reliability data. He is a three-time recipient of the Frank Wilcoxon prize for the best practical application paper in *Technometrics*, a four-time recipient of the W.J. Youden prize for the best expository paper in *Technometrics*, and a recipient of the Shewhart Medal for outstanding technical leadership in the field of modern quality control. He is a fellow of the American Statistical Association, an elected member of the International Statistical Institute, and a fellow of the American Society for Quality. He received his Ph.D. in administrative and engineering systems from Union College.

NACHI NAGAPPAN works on empirical software engineering and measurement at Microsoft Research. His research interests are in the fields of software reliability, software measurement and testing, and empirical software engineering. He has also worked on social factors in software engineering, aspect-oriented software development, and computer science education. His current research focuses on the application of software measurement and statistical modeling to large software systems and the next-generation Windows operating system (Vista). He holds a Ph.D. from North Carolina State University.

MICHAEL PECHT is George E. Dieter professor of mechanical engineering and a professor of applied mathematics at the University of Maryland. At the university, he was the founder of the Center for Advanced Life Cycle Engineering, which is funded by more than 150 of the world's leading electronics companies. His main research interest is the development of reliable electronics products and use and supply chain management. He is a licensed professional engineer in the state of Maryland. He is a fellow of the Institute of Electrical and Electronics Engineers, a fellow of ASME, a fellow of the Society of Automotive Engineers, and a fellow of the International Microelectronics Assembly and Packaging Society. He is a recipient of the exceptional technical achievement award and the lifetime achievement award of the Institute of Electrical and Electronics Engineers. He holds a Ph.D. from the University of Wisconsin–Madison.

ANANDA SEN is an associate research scientist at the Center for Statistical Consultation and Research at the University of Michigan. Previously, he held teaching appointments at Oakland University and the University of Michigan. His primary work focuses on the understanding of reliability growth modeling, accelerated failure-time modeling, and Bayesian methodologies in reliability and survival analysis. He is a fellow of the American Statistical Association and an elected member of the International Statistical Institute. He holds an M.S. in statistics from the Indian Statistical Institute and a Ph.D. in statistics from the University of Wisconsin–Madison.

SCOTT VANDER WIEL is technical staff member at Los Alamos National Laboratory. Previously, he conducted statistics research at Bell Laboratory. In his work, he collaborates with engineers and scientists to analyze data and develop statistical methodology in system reliability (and other areas of the application of statistics). At Los Alamos, he has focused on weapons reliability modeling and uncertainty quantification. He is a fellow of the American Statistical Association. He holds an M.A. and a Ph.D. in statistics from Iowa State University.

COMMITTEE ON NATIONAL STATISTICS

The Committee on National Statistics was established in 1972 at the National Academies to improve the statistical methods and information on which public policy decisions are based. The committee carries out studies, workshops, and other activities to foster better measures and fuller understanding of the economy, the environment, public health, crime, education, immigration, poverty, welfare, and other public policy issues. It also evaluates ongoing statistical programs and tracks the statistical policy and coordinating activities of the federal government, serving a unique role at the intersection of statistics and public policy. The committee's work is supported by a consortium of federal agencies through a National Science Foundation grant.